I0064617

Introduction to *2-Spinors* in General Relativity

Introduction to *2-Spinors* in General Relativity

Peter O'Donnell
Cambridge University, UK

World Scientific
NEW JERSEY • LONDON • SINGAPORE • BEIJING • SHANGHAI • HONG KONG • TAIPEI • CHENNAI • TOKYO

Published by

World Scientific Publishing Co. Pte. Ltd.
5 Toh Tuck Link, Singapore 596224
USA office: 27 Warren Street, Suite 401-402, Hackensack, NJ 07601
UK office: 57 Shelton Street, Covent Garden, London WC2H 9HE

British Library Cataloguing-in-Publication Data
A catalogue record for this book is available from the British Library.

INTRODUCTION TO 2-SPINORS IN GENERAL RELATIVITY

Copyright © 2003 by World Scientific Publishing Co. Pte. Ltd.

All rights reserved. This book, or parts thereof, may not be reproduced in any form or by any means, electronic or mechanical, including photocopying, recording or any information storage and retrieval system now known or to be invented, without written permission from the publisher.

For photocopying of material in this volume, please pay a copying fee through the Copyright Clearance Center, Inc., 222 Rosewood Drive, Danvers, MA 01923, USA. In this case permission to photocopy is not required from the publisher.

ISBN-13 978-981-238-307-5
ISBN-10 981-238-307-7

For My Grandparents

Preface

This book is an introduction to the two-component spinor, or 2-spinor, formalism in general relativity. It has been designed to be readily accessible to those at graduate and research level.

The general mathematical framework of spinors was first introduced by the French mathematician Élie Cartan (1869–1951). Now some ninety years after its conception, spinors are used extensively in the fields of general relativity, quantum mechanics, particle physics, superstrings and M-Theory. Most theoretical physicists are familiar with 4-spinors due to the work of Dirac, and those involved in ten-dimensional superstring theory frequently employ 32-spinors. However, it is primarily through the work of Penrose that the 2-spinor formalism is widely used today in general relativity. Many of the notations and conventions introduced by Penrose are adhered to in this book; particularly those given in Penrose and Rindler, 1984, 1986. Part of the intention here is to equip the reader with the necessary information and techniques to enable him or her to progress to more detailed expositions on the subject, such as the material covered in the aforementioned text and appropriate research articles.

Chapter 1 begins by developing the 2-spinor formalism in a geometrical fashion. This is carried out by associating the real Minkowski coordinates on a Riemann sphere with a single complex one on a complex 2-plane via stereographical projections. A one-to-one correspondence can thereby be established between coordinate systems provided that each single complex 'point' is defined in terms of a complex 2-form. The elements of the 2-form will then be seen to form an orthonormal basis in spin-space. Each point in spin space can then be associated with a geometrical object called a spin-vector — the simplest kind of 2-spinor.

In Chapter 2 we build the spinor algebra and construct more general spinors. The connection between spinors and tensors is established through the use of Infeld–van der Waerden symbols. Furthermore, the decomposition of certain bivalent tensors into their spinor equivalents is discussed. Chapter 2 ends with the algebraic classification of the electromagnetic and Weyl spinors.

Chapter 3 commences with the axiomatic description of the spinor covariant derivative. We go on to introduce the spinor affine connections (Ricci rotation coefficients), curvature spinor and associated spinor fields. This culminates in a discussion of the Newman–Penrose spin coefficient formalism.

The main theme of Chapter 4 is the Lanczos spinor and the Weyl–Lanczos equations. The origins of this spinor are first highlighted together with its significance in general relativity. The Weyl–Lanczos equations are derived (using techniques given in Chapters 2 and 3) and some solutions are obtained.

For the benefit of advanced undergraduates who have done little or no general relativity, but nevertheless wish to gain some knowledge of the 2-spinor formalism, an appendix has been included to review the fundamentals of general relativity.

A number of selected exercises have been given at the end of each chapter.

Finally, I would like to thank Dr. Jennifer Atkinson for creating the index, proofreading and typing the manuscript; Professor Sir Roger Penrose for his support for this work; and Ian Seldrup and Anthony Doyle at World Scientific for their help during the preparation of the manuscript.

Peter O'Donnell
Cambridge
U.K.

Contents

Chapter 1

Spinor geometry

It has been argued that the most natural way of discussing *spinors* is by means of the *theory of representations of groups* (see, for example, Cartan, 1966; Carmeli and Malin, 2000). Indeed, most of the literature currently available on spinors follows this precedent and develops the theory exceedingly well. However, we are interested in spinors applied to general relativity theory (a geometrical theory) and, in this sense, a geometrical approach to studying spinors will be the most appropriate one.

1.1 Minkowski space

We define *Minkowski space* $\mathbb{M}$ to be a four-dimensional vector space over the real number field $\mathbb{R}$ with a flat *Lorentzian metric* $\eta_{\mathbf{ab}}$ given by

$$\eta_{\mathbf{ab}} = diag(1, -1, -1, -1). \tag{1.1}$$

(Note that boldface indices will always represent numerical values: lower-case Latin indices will range over $0, 1, 2, 3$, and upper-case Latin indices will range over $0, 1$.) At each point on $\mathbb{R}^4$ there exists a set of *basis* vectors $\mathbf{e}_0, \mathbf{e}_1, \mathbf{e}_2, \mathbf{e}_3, \in \mathbb{M}$ called a *tetrad*, which define uniquely any $\mathbf{U} \in \mathbb{M}$ by

$$\mathbf{U} = U^0\mathbf{e}_0 + U^1\mathbf{e}_1 + U^2\mathbf{e}_2 + U^3\mathbf{e}_3 \tag{1.2}$$

for $U^0, U^1, U^2, U^3 \in \mathbb{R}$ not all zero unless $\mathbf{U} = \mathbf{0}$. We can write (1.2) more concisely as

$$\mathbf{U} = U^{\mathbf{a}}\mathbf{e}_{\mathbf{a}} \tag{1.3}$$

where it is understood that from now, and henceforth, the *Einstein summation convention* will be adopted whenever indices are repeated as in (1.3).

An *inner product (scalar product)* on $\mathbb{M}$ is a mapping $\mathbb{M} \times \mathbb{M} \to \mathbb{R}$. Thus for any $\mathbf{U}$, $\mathbf{V}$, $\mathbf{W}$, $\in \mathbb{M}$ and $a, b \in \mathbb{R}$ we have

$$\mathbf{U} \cdot \mathbf{V} = \mathbf{V} \cdot \mathbf{U} \tag{1.4}$$

$$\mathbf{U} \cdot (a\mathbf{V} + b\mathbf{W}) = a\mathbf{U} \cdot \mathbf{V} + b\mathbf{U} \cdot \mathbf{W} \tag{1.5}$$

$$\mathbf{U} \cdot \mathbf{V} = 0 \quad \forall \mathbf{V} \in \mathbb{M} \Leftrightarrow \mathbf{U} = 0. \tag{1.6}$$

Hence the inner product is *symmetric* by (1.4) and *bilinear* from (1.4) and (1.5). Equation (1.6) implies that the inner product is also *non-degenerate*, and we say that $\mathbb{M}$ is *non-singular*. The *orthonormalisation* conditions on the tetrad $\mathbf{e_a}$ are

$$\mathbf{e_a} \cdot \mathbf{e_b} = \begin{cases} 1 & \text{if } \mathbf{a} = \mathbf{b} = 0 \\ -1 & \text{if } \mathbf{a} = \mathbf{b},\ \mathbf{a} = 1, 2, 3 \\ 0 & \text{if } \mathbf{a} \neq \mathbf{b} \end{cases} \tag{1.7}$$

which can be written compactly as

$$\mathbf{e_a} \cdot \mathbf{e_b} = \eta_{\mathbf{ab}} \tag{1.8}$$

where $\eta_{\mathbf{ab}} = \eta^{\mathbf{ab}}$ is given by (1.1).

We shall adopt the convention of referring to a tetrad $\mathbf{e_a}$ which satisfies (1.8) as a *Minkowski tetrad*, and coordinates $U^{\mathbf{a}} \in \mathbb{R}$ representing any $\mathbf{U} \in \mathbb{M}$ will be called *Minkowski coordinates*.

A vector $\mathbf{U} \in \mathbb{M}$ is called

$$\left.\begin{array}{ll} \text{timelike} & \text{if } \mathbf{U} \cdot \mathbf{U} > 0 \\ \text{null} & \text{if } \mathbf{U} \cdot \mathbf{U} = 0 \\ \text{spacelike} & \text{if } \mathbf{U} \cdot \mathbf{U} < 0. \end{array}\right\} \tag{1.9}$$

In Minkowski coordinates $\mathbf{U}$ is timelike or null if

$$\mathbf{U} \cdot \mathbf{U} = (U^0)^2 - (U^1)^2 - (U^2)^2 - (U^3)^2 \geq 0. \tag{1.10}$$

It can be easily shown that a timelike vector cannot be orthogonal to a null vector (see Exercise 1.1) or, indeed, to another timelike vector. Because of this, timelike vectors and null vectors can be separated into two classes i.e. *future-pointing* and *past-pointing*, then $\mathbb{M}$ is said to be *time-orientated*. We call the Minkowski tetrad $\mathbf{e_a}$ *orthochronous* if $\mathbf{e_0}$ is a future-pointing timelike vector, or more succinctly, a *future-timelike vector*.

Consider two Minkowski tetrads $\mathbf{e_a}$ and $\mathbf{e_{a'}}$. Each can be related to the other by a general linear combination of vectors written compactly as

$$\mathbf{e_a} = \Lambda_{\mathbf{a}}{}^{\mathbf{a}'} \mathbf{e_{a'}} \tag{1.11}$$

where $(\Lambda_{\mathbf{a}}{}^{\mathbf{a}'})$ is a transformation matrix that is real and non-singular. Consequently, the determinant of this matrix is non-zero. If det $(\Lambda_{\mathbf{a}}{}^{\mathbf{a}'}) > 0$ then $\mathbf{e_a}$ and $\mathbf{e_{a'}}$ are defined to be *equally orientated*, and thus the tetrads are *proper*. If det $(\Lambda_{\mathbf{a}}{}^{\mathbf{a}'}) < 0$ then $\mathbf{e_a}$ and $\mathbf{e_{a'}}$ are defined to be *unequally orientated*, and the tetrads are *improper*. $\mathbb{M}$ is said to be *orientated* depending on whether $\mathbf{e_a}$ is proper or improper.

It will be assumed throughout that our orthonormal basis will be *right-handed*, that is, $\mathbf{e_a}$ is proper in addition to being orthochronous — otherwise $\mathbf{e_a}$ would be deemed to be *left-handed.* These two requirements, taken together, also form what is called a *restricted Minkowski tetrad.* (Note that if $\mathbf{e_a}$ was neither proper nor orthochronous then our basis would still be right-handed, but it would no longer be restricted.)

We define a *Lorentz transformation* to be *restricted* if an *active Lorentz transformation* is both orientated and time-orientated on $\mathbb{M}$. The adjective 'active' refers to a linear transformation of $\mathbb{M}$ such that the first equality of (1.10) is presumed. A *passive Lorentz transformation* is a map

$$P : U^{\mathbf{a}} \mapsto U^{\mathbf{a}'} \tag{1.12}$$

where $U^{\mathbf{a}} \in \mathbb{R}$ are Minkowski coordinates defined by some Minkowski tetrad $\mathbf{e_a} \in \mathbb{M}$. If the tetrads that form (1.12) are themselves restricted then (1.12) is *restricted.* If the two tetrads in question are given by (1.11) we can write (1.12) as

$$U^{\mathbf{a}'} = U^{\mathbf{a}} \Lambda_{\mathbf{a}}{}^{\mathbf{a}'}. \tag{1.13}$$

Although active Lorentz transformations are coordinate independent it is still permissible to discuss them in terms of coordinates. To do this, we define an active Lorentz transformation as a map

$$A : U^{\mathbf{a}} \mapsto V^{\mathbf{a}'}. \tag{1.14}$$

If also $\mathbf{e_a} \mapsto \mathbf{e_{a'}} \in \mathbb{M}$ then we can write $U^{\mathbf{a}} = V^{\mathbf{a}'}$. Thus the passive transformation generated by $\mathbf{e_{a'}} \mapsto \mathbf{e_a}$ sends $U^{\mathbf{a}}$ to $V^{\mathbf{a}}$. With the aid of (1.13) we arrive at

$$U^{\mathbf{a}'} = V^{\mathbf{b}'} \Lambda_{\mathbf{b}}{}^{\mathbf{a}'} \tag{1.15}$$

where in this particular case, summation is over $\mathbf{b}'$ and $\mathbf{b}$. Thus the active Lorentz transformation is given by

$$V^{\mathbf{b}'} = U^{\mathbf{a}'} A_{\mathbf{a}'}{}^{\mathbf{b}} \tag{1.16}$$

where the matrix

$$(A_{\mathbf{a}'}{}^{\mathbf{b}}) = (\Lambda_{\mathbf{b}}{}^{\mathbf{a}'})^{-1}. \tag{1.17}$$

1.2 The null cone and Riemann sphere

Our aim is to establish a coordinate representation of what will come to be known as a *spin-vector*; that is, the object which is considered to be the simplest of the generic class of spinors.

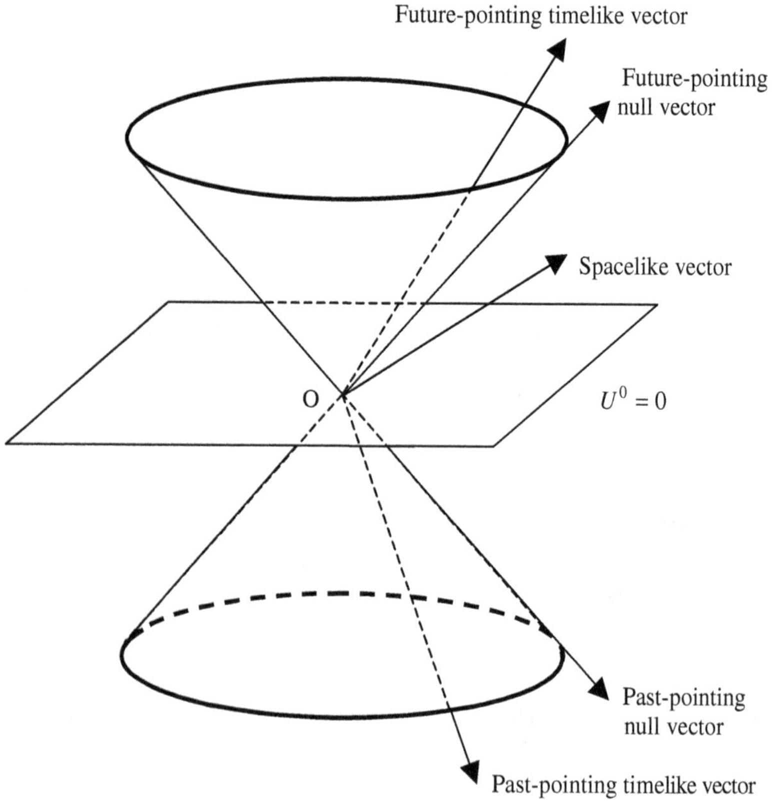

Fig. 1.1 The null cone

The set of all null vectors forms a null cone (see Fig. 1.1) with coordinates represented by

$$(U^0)^2 - (U^1)^2 - (U^2)^2 - (U^3)^2 = 0. \tag{1.18}$$

For any timelike or null vector $\mathbf{U} \in \mathbb{M}$ given by (1.2) we associate two distinct and opposite *directions* relative to the origin O; namely, *future-timelike/null directions* and *past-timelike/null directions.* This will be elucidated upon presently.

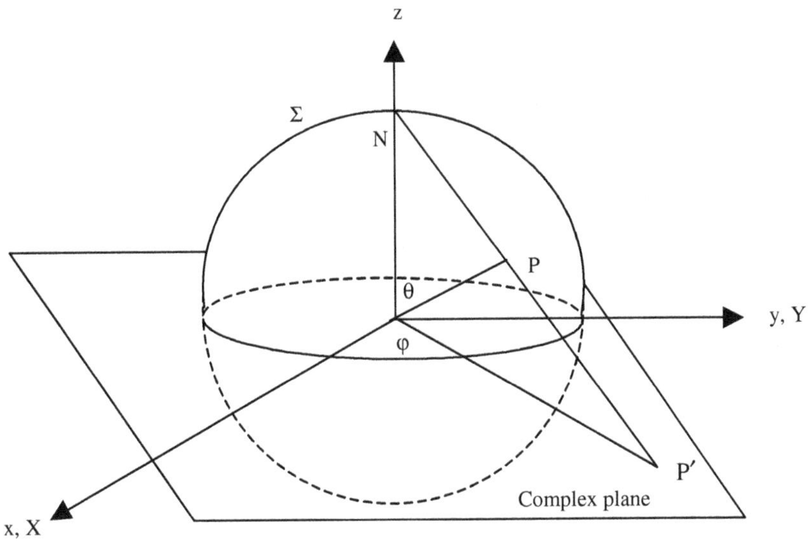

Fig. 1.2 Stereographic projection of the Riemann sphere

Consider the null cone as consisting of two distinct, but connected, halves separated by the hyperplane $U^0 = 0$, the upper part being the *future null cone* and the lower half being the *past null cone*. Furthermore, we can associate with each half of the null cone an abstract space with components that are either future null directions or past null directions. If we now form an intersection with, say, the past null cone and some hyperplane $T =$ constant then topologically, this would represent a sphere in $\mathbb{R}^3$. Indeed, in the space-time of special relativity, the observer at the origin would consider this sphere as his entire field of vision, i.e. his *celestial sphere*. Hence null rays would constitute points on the celestial sphere. Points on the hyperplane T = constant define the direction of any vector given by (1.2), provided $T \neq 0$. Timelike directions are given by points inside the sphere and spacelike directions are given by points outside.

Consider now the *anti-celestial sphere*, i.e. the abstract space representing a sphere in $\mathbb{R}^3$ when the future null cone is intersected by the hyperplane $T = 1$, for example. Mathematically, we might think of this as the *Riemann sphere*, Σ, given by the equation

$$x^2 + y^2 + z^2 = 1. \tag{1.19}$$

By means of *stereographic projection* a one-to-one correspondence can be set up between the *extended complex plane* $\tilde{\mathbb{C}}$ and Σ (see Fig. 1.2). The coordinates x, y, z on Σ can then be replaced with a single complex parameter on the complex plane $\mathbb{C}$. Clearly $\tilde{\mathbb{C}}$ is formed by adding to $\mathbb{C}$ an extra 'point' ∞ — which represents the 'north pole' of Σ.

The geometrical construction is as follows. $\mathbb{C}$ intersects Σ at the 'equator' $z = 0$ where we think of $\mathbb{C}$ as being embedded in Euclidean space $\mathbb{R}^3$, $T = 1$. Each point on Σ is stereographically projected, or mapped, from the north pole, N, with coordinates $(1, 0, 0, 1)$, to $\mathbb{C}$. Thus the mapping carries the point P, with coordinates $(1, x, y, z)$, to a point P' with coordinates $(1, X, Y, 0)$. Each point on $\mathbb{C}$ can then be identified by a single complex number, sometimes called a *stereographic coordinate*

$$\xi = X + iY. \tag{1.20}$$

Using simple trigonometry we can see that

$$\frac{NP}{NP'} = \frac{x}{X} = \frac{y}{Y} = 1 - z. \tag{1.21}$$

Comparing (1.21) with (1.20) implies that

$$\xi = \frac{x + iy}{1 - z}. \tag{1.22}$$

In spherical polar coordinates (θ, ϕ) equation (1.22) can be written as

$$\xi = e^{i\phi} \cos \frac{\theta}{2}, \tag{1.23}$$

which is achieved by applying the standard equations

$$x = \sin\theta \cos\phi, \quad y = \sin\theta \sin\phi, \quad z = \cos\theta \tag{1.24}$$

to (1.22). The inverse relations with respect to (1.22) are directly obtainable with the aid of (1.19). Thence,

$$x = \frac{\xi + \bar{\xi}}{\xi\bar{\xi} + 1}, \quad y = \frac{i(\bar{\xi} - \xi)}{\xi\bar{\xi} + 1}, \quad z = \frac{\xi\bar{\xi} - 1}{\xi\bar{\xi} + 1}. \tag{1.25}$$

1.3 Spin transformations and spin matrices

It will be preferable to represent the single complex parameter, ξ, by a pair of complex components

$$\xi = \frac{\zeta}{\eta}. \tag{1.26}$$

The reason for doing this is so as to circumvent the problem of using an infinite coordinate to represent the north pole on the Riemann sphere. (It will be shown later that the pair (ζ, η) can be treated as components of a spin-vector.) With the complex components (1.26), (1.25) can be rewritten as

$$x = \frac{\zeta\bar{\eta} + \bar{\zeta}\eta}{\zeta\bar{\zeta} + \eta\bar{\eta}}, \quad y = \frac{i(\bar{\zeta}\eta - \zeta\bar{\eta})}{\zeta\bar{\zeta} + \eta\bar{\eta}}, \quad z = \frac{\zeta\bar{\zeta} - \eta\bar{\eta}}{\zeta\bar{\zeta} + \eta\bar{\eta}}. \tag{1.27}$$

If (1.27) is multiplied by $\zeta\bar{\zeta} + \eta\bar{\eta}$ — the denominator of (1.27) — then any point on the null cone with coordinates (T, X, Y, Z) can be represented in terms of the complex pair (ζ, η) by

$$T = \frac{1}{\sqrt{2}}(\zeta\bar{\zeta} + \eta\bar{\eta}), \quad X = \frac{1}{\sqrt{2}}(\zeta\bar{\eta} + \eta\bar{\zeta})$$
$$Y = \frac{i}{\sqrt{2}}(\bar{\zeta}\eta - \eta\bar{\zeta}), \quad Z = \frac{1}{\sqrt{2}}(\zeta\bar{\zeta} - \eta\bar{\eta}). \tag{1.28}$$

Notice that the factor $\frac{1}{\sqrt{2}}$ has been introduced so as to be consistent with analogous expressions to be given in the next chapter.

Let

$$\zeta \mapsto \hat{\zeta} = a\zeta + b\eta$$
$$\eta \mapsto \hat{\eta} = c\zeta + d\eta \tag{1.29}$$

be a general complex linear transformation of components ζ and η, where $a, b, c, d \in \mathbb{C}$ and $ad - bc \neq 0$. Hence a conformal transformation of the Riemann sphere must be a globally defined holomorphic transformation

$$\xi \mapsto \hat{\xi} = \frac{a\xi + b}{c\xi + d}. \tag{1.30}$$

Without loss of generality we may take $ad - bc = 1$, then this condition together with (1.30) (or independently (1.29)), are referred to as *spin transformations.*

The non-singular matrix $\mathbf{S}$ defined by

$$\mathbf{S} = \begin{pmatrix} a\, b \\ c\, d \end{pmatrix}, \quad ad - bc = 1 \tag{1.31}$$

is called the *spin matrix.* Thus (1.29) can be rewritten as

$$\begin{pmatrix} \hat{\zeta} \\ \hat{\eta} \end{pmatrix} = \mathbf{S} \begin{pmatrix} \zeta \\ \eta \end{pmatrix}. \tag{1.32}$$

Clearly, a double application of (1.32) yields yet another spin transformation. $\mathbf{S}$ being non-singular ensures that $\mathbf{S}^{-1}$ exists. And a group is indicated; this group is called the *special linear group*, commonly referred to as $SL(2, \mathbb{C})$.

1.4 Flagpoles and flag planes

Let us suppose that (1.28) is the coordinate representation of a future-pointing null vector $\mathbf{A}$ defined by the directed line segment $\overrightarrow{OQ}$ on the future null cone (see Fig. 1.3). In this context $\mathbf{A}$ is referred to as a 'flagpole'. On the Riemann sphere itself $\mathbf{A}$ has coordinates (ζ, η). Clearly, if we refer to (1.28), $\mathbf{A}$ remains invariant under a *transformation of phase* $\zeta \mapsto e^{i\theta}\zeta$, $\eta \mapsto e^{i\theta}\eta$, and consequently provided θ is real the same null vector is defined.

If we now introduce a real spacelike vector $\mathbf{B}$, which is a tangent vector to the Riemann sphere and orthogonal to $\mathbf{A}$, then each positive multiple of $\mathbf{B}$ lies in a two-plane $s\mathbf{A} + t\mathbf{B}, (s, t \in \mathbb{R})$. This plane is tangent to the future null cone because $\mathbf{A} \cdot \mathbf{B} = 0$.

In actuality, $s\mathbf{A} + t\mathbf{B}$ represents a *half-plane* for $t > 0$. This is required if we wish to discuss the 'orientation' of $\mathbf{B}$ (not to be confused with our previous use of orientation). This half-plane is our 'flag plane'. Combining the flagpole and flag plane forms a structure known as a *null flag.*

Consider the spin transformation (1.32) with $\mathbf{S} = diag(e^{i\theta}, e^{i\theta})$:

$$\begin{pmatrix} \hat{\zeta} \\ \hat{\eta} \end{pmatrix} = \begin{pmatrix} e^{i\theta} & 0 \\ 0 & e^{i\theta} \end{pmatrix} \begin{pmatrix} \zeta \\ \eta \end{pmatrix}. \tag{1.33}$$

For $0 \leq \theta \leq \pi$ we have

$$\begin{pmatrix} \hat{\zeta} \\ \hat{\eta} \end{pmatrix} = - \begin{pmatrix} \zeta \\ \eta \end{pmatrix}, \tag{1.34}$$

however, the flag plane has performed a 2π rotation! That is: a phase change of π leaves the flag plane invariant and takes $\mathbf{S}$ to $-\mathbf{I}$ (where $\mathbf{I}$ is the identity matrix).

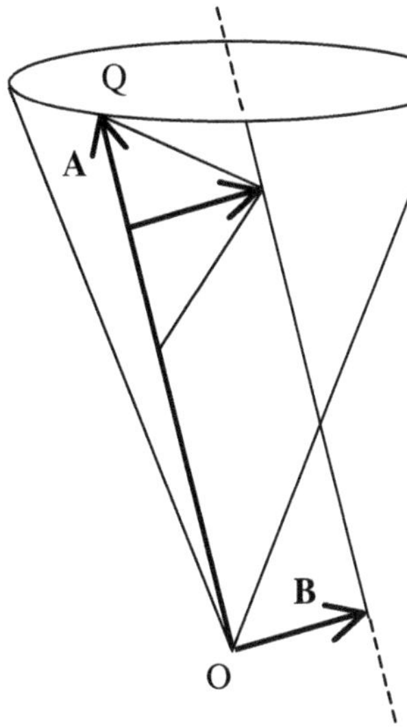

Fig. 1.3 The spin-vector as a null flag

For $\pi \leq \theta \leq 2\pi$ (1.33) becomes

$$\begin{pmatrix} \hat{\zeta} \\ \hat{\eta} \end{pmatrix} = \begin{pmatrix} \zeta \\ \eta \end{pmatrix}. \tag{1.35}$$

Thus a further 2π rotation must necessarily be performed by the flag plane in order that the null flag is returned to its original state.

A glance at (1.34) and (1.35) shows clearly that a sign ambiguity exists in the local geometry of $\mathbb{M}$; an ambiguity which cannot be resolved utilising geometrical reasoning. Indeed, the scope of local geometry must be extended to allow for the existence of *spinorial objects*. We will not enter into a detailed discussion here regarding these objects, and the reader is referred to the appropriate literature in the bibliography. However, roughly speaking, a null flag can be regarded as comprising of two spin-vectors — the spin-vector itself being a spinorial object. For example, consider the two arbitrary vectors $\boldsymbol{\alpha}$ and -$\boldsymbol{\alpha}$ defined by some null flag. We can send $\boldsymbol{\alpha}$ into -$\boldsymbol{\alpha}$ by performing a continuous 2π rotation. A further 2π rotation results in the original spin-vector. (It is of some interest to mention that an analogous description of flags has been given by Payne, 1952. In that paper flags are axes, the flag plane is the blade of the axe and the flagpole is the axe handle.)

1.5 Spin-space

Spin-vectors $\boldsymbol{\zeta}$, $\boldsymbol{\eta}$, etc. are elements belonging to *spin-space* S. We define S to be a two-dimensional vector space over the complex field $\mathbb{C}$. Furthermore, there exists on S a two-form $[\,,\,]$ with the following properties. Namely,

(1) a *skew-symmetric* (*alternating*) inner product, which is

(2) *bilinear*, and

(3) *non-degenerate.*

Let us consider these properties in more detail. Property (1) can be represented in a straightforward manner by

$$[\boldsymbol{\zeta}, \boldsymbol{\eta}] = -[\boldsymbol{\eta}, \boldsymbol{\zeta}] \tag{1.36}$$

for all $\boldsymbol{\zeta}$, $\boldsymbol{\eta} \in S$ and $[\boldsymbol{\zeta}, \boldsymbol{\eta}] \in \mathbb{C}$. Property (2) is directly represented by the relations:

$$\lambda[\boldsymbol{\zeta}, \boldsymbol{\eta}] = [\lambda\boldsymbol{\eta}, \boldsymbol{\zeta}] \tag{1.37}$$

$$[\boldsymbol{\zeta} + \boldsymbol{\eta}, \boldsymbol{\phi}] = [\boldsymbol{\zeta}, \boldsymbol{\phi}] + [\boldsymbol{\eta}, \boldsymbol{\phi}] \tag{1.38}$$

for $\lambda \in \mathbb{C}$. In combination, relations (1.37) and (1.38) imply that $[\,,\,]$ is linear in the second argument, and (1.36) ensures linearity in the first argument. Hence bilinearity ensues.

It is evident that for two linearly dependent spin-vectors $\boldsymbol{\kappa} = \lambda\boldsymbol{\zeta}$, say, it immediately follows from (1.36) and (1.37) that

$$[\boldsymbol{\zeta}, \boldsymbol{\kappa}] = 0. \tag{1.39}$$

Therefore a necessary and sufficient condition for two spin-vectors to be linearly dependent is

$$[\boldsymbol{\zeta}, \boldsymbol{\eta}] = 0 \Rightarrow \boldsymbol{\zeta} = \mathbf{0} \tag{1.40}$$

for all $\boldsymbol{\eta} \in S$, or

$$[\boldsymbol{\zeta}, \boldsymbol{\eta}] \neq 0 \tag{1.41}$$

for all $\boldsymbol{\zeta}$, $\boldsymbol{\eta} \in S$. This then is the explicit form of the non-degeneracy property (3). Notice the set $\{\boldsymbol{\zeta}, \boldsymbol{\eta}\}$ of spin-vectors (not proportional to each other) forms a spanning set for S.

Let $(\boldsymbol{o}, \boldsymbol{\iota})$ constitute a *spin basis* for S where $\boldsymbol{o}$ and $\boldsymbol{\iota}$ are two arbitrary spin-vectors such that

$$[\boldsymbol{o}, \boldsymbol{\iota}] = 1 = -[\boldsymbol{\iota}, \boldsymbol{o}] \tag{1.42}$$

by (1.36). Then we define the components of some spin-vector $\boldsymbol{\zeta} \in S$ by

$$\boldsymbol{\zeta} = \zeta^0 \boldsymbol{o} + \zeta^1 \boldsymbol{\iota} \tag{1.43}$$

where

$$\zeta^0 = [\boldsymbol{\zeta}, \boldsymbol{\iota}], \qquad \zeta^1 = -[\boldsymbol{\zeta}, \boldsymbol{o}] \tag{1.44}$$

and

$$\boldsymbol{o} = (1,0), \qquad \boldsymbol{\iota} = (0,1). \tag{1.45}$$

Remark: in this section, as in the whole of this book, we have adopted the convention that both spin-vector and its corresponding components are represented by the same Greek kernel letter. No ambiguity will result with regard to previous sections. All that has taken place is a relabelling of components i.e. in the case of (1.43) $\zeta \to \zeta^0$ and $\eta \to \zeta^1$.

By direct calculation one can easily show that the component form of the inner product (1.36) is

$$[\boldsymbol{\zeta}, \boldsymbol{\eta}] = \zeta^0 \eta^1 - \zeta^1 \eta^0 \tag{1.46}$$

where (η^0, η^1) are the components of $\boldsymbol{\eta}$.

This result will be useful in our study of the *Levi–Civita spinor* — which will be defined later. Before we continue with our explanation of spin bases, it will be advantageous to discuss what has come to be known as 'abstract index notation' (Penrose, 1968; Penrose and Rindler, 1986). This deals with the inherent ambiguities involving indices.

1.6 Exercises

1.1 Let $\mathbf{T}$ be a timelike vector with components $(T, 0, 0, 0)$ and $\mathbf{V}$ be a vector orthogonal to $\mathbf{T}$. Relative to an orthonormal basis in Minkowski space show that a timelike vector cannot be orthogonal to a null vector.

1.2 Obtain equation (1.13).

1.3 Obtain equation (1.23) by applying trigonometry to Fig. 1.2.

1.4 Using the *antipodal transformation* $(x, y, z) \mapsto (-x, -y, -z)$ and the equivalent transformation for polar coordinates $(\theta, \phi) \mapsto (\pi - \theta, \pi + \phi)$ obtain the antipodal counterparts of equations (1.22), (1.23), and (1.25).

[Hint: you should find in all cases that the effect of these transformations is $\xi \mapsto -\overline{\xi}^{-1}$.]

1.5 What geometrical inference can be made regarding the above antipodal transformations?

1.6 A child's swing is fixed to the branch of a tree by looping two pieces of rope over the branch and connecting the free ends to each of the four corners of the swing's seat. The seat is then rotated 720° about an axis through the centre of the seat causing the rope to become twisted. How can the rope be disentangled without rotating the seat again? Note that chopping down the tree is not an option!

[Remark: the above exercise is a variation of Paul Dirac's scissors problem. Other variations can also be found (see Penrose and Rindler, 1984; Misner, Thorne and Wheeler, 1973; Huggett and Tod, 1985).]

1.7 Verify (1.46).

Chapter 2

Spinor algebra

In the last chapter we discussed the concept of a spin-vector with respect to its purely geometrical characterisation. This necessitated the view of associating the structure of a spin-vector with that of a null flag in an analogous way whereby one can represent a Cartesian vector by a directed line segment. Moreover, rotating the spin-vector by 2π reverses its sign, and only by a further 2π rotation does the spin-vector return to its original orientation.

The geometrical interpretation of a spin-vector and its associated spin-space was primarily intended to aid the visualisation of what is in many ways an abstract entity. In fact further development does not require the same kind of conceptualised geometrical interpretation.

As before our spin-space S will consist of spin-vectors at a point in space-time, then S is the spin-space at that particular point. However, the properties between spin-vectors at a point can be carried over to spin-vector fields on space-time in a straightforward manner.

2.1 Abstract index notation

So far our analysis has been in the main geometrically based. And the geometrical objects under discussion (spin-vectors) have been treated in an effectively component independent fashion. Indeed, it is considered by many mathematicians, and even by some physicists, that calculations which are littered with vast arrays of tiny subscripts and superscripts can tend to obscure their true meaning. For example, with respect to a quantity V_a, do we mean a vector V_a or a vector which contains a component V_a? Furthermore, one invariably feels the necessity to introduce a coordinate system when performing computations using vectors or tensors. However, many contemporary descriptions of a tensor are in fact unrelated to coordinate systems, and even the concept of a coordinate system can be avoided.

It appears then that we can dispense totally with the obligation of endowing tensor calculations with these troublesome components. Unfortunately, this is not the case. Almost all tensored operations can be performed in a component independent way, i.e. addition, multiplication, contraction etc. with the possible exception of index permutation. For all but the simplest of these operations, introducing components into the notation appears to be the only satisfactory course. (It is of some interest to be aware of certain diagrammatical interpretations of tensor and spinor operations, (Penrose, 1971; Cvitanovic, 1976; Cvitanovic and Kennedy, 1982.) These interpretations have circumvented the requirement of adopting indices during all tensor operations, including that of permutation. We will not discuss the merits of such a notation here, and the interested reader is referred to the appropriate references in the bibliography.) Thus, the usefulness of indices is apparent and we will employ them throughout.

Consider the geometrical object ζ which we introduced in the last chapter and have referred to as a spin-vector. One may be compelled to associate with ζ an object ζ^A, and think of it as representing an ordered set of components of ζ where the index would commonly take integer values. However, we must dissociate ourselves from this compulsion, and instead treat ζ^A as a spin-vector in its own right. In fact ζ^A is just another way of writing ζ, only now we have labelled it with the *abstract index* A. Of course, we would at some time need to identify the components of ζ^A with respect to a basis. This can be achieved via the object $\zeta^{\mathbf{A}}$ where now $\mathbf{A}$ ranges over $0, 1$ because of the two-dimensionality of spin-vectors. (Note that by convention $0, 1$ is used for component indices rather than $1, 2$ in some older texts. This is mainly due to the former convention's visual similarity with the Greek letters o, ι which formed our spin-frame in the last chapter.)

Let us now introduce another spin-vector distinct from ζ^A but nevertheless representing the spin-vector ζ, and label it with the abstract index $B : \zeta^B$. Now ζ^A and ζ^B must be distinct for if they were proportional to each other then relationships like $\zeta^A = \zeta^B$ would exist. Such relationships are impossible in standard tensor theory, and so must it also be in spinor theory. It is clear then that for an arbitrary spin-vector ζ we must construct an infinite, but distinct, set of isomorphic copies labelled ζ^A, ζ^B,.... Furthermore, since ζ is a an element of S there must also be an infinite, but distinct, set of isomorphic copies of S labelled S^A, S^B,....

Although not explicitly stated the objects ζ^A, ζ^B are spinors, or more precisely, univalent spinors or $(1,0;0,0)$ spinors. The $(1,0;0,0)$ is called the spinor *valence* which, with respect to our discussion so far, amounts to a single contravariant index. More about higher or multivalent spinors presently.

We denote the dual space of S by S^*. For this space we also construct an infinite, but distinct, set of isomorphic copies of S^* labelled S_A, $S_B, \ldots$. The element of the dual spaces $\zeta_A \in S_A$, $\zeta_B \in S_B, \ldots$ are isomorphic to each other.

For each element $\zeta \in S$ we can identify $[\zeta, \,] \in S^*$ such that $[\zeta, \,]$ is a linear map $[\zeta, \,] : S \to \mathbb{C}$ i.e. for $\zeta_A \in S_A$, ζ_A is a linear map $\zeta_A : S^A \to \mathbb{C}$ then $\zeta_A \eta^A \in \mathbb{C}$ is called an *inner product* or *contraction* or *transvection*. The same Greek kernel letter, ζ, will be used for elements of S^A and the dual S_A.

A univalent spinor ζ_A with a single covariant index is a $(0,0;1,0)$ spinor i.e. a spinor with valence $(0,0;1,0)$. We can construct higher valence spinors by analogy with the above. Thus, a $(p,0;r,0)$ spinor $\chi^{A\ldots C}_{D\ldots F}$ with p contravariant indices and r covariant indices is a bilinear map $\chi^{A\ldots C}_{D\ldots F} : S_A \times \cdots \times S_C \times S^D \times \cdots \times S^F \to \mathbb{C}$ where $\chi^{A\ldots C}_{D\ldots F}$ is an element of $S^{A\ldots C}_{D\ldots F}$. Indeed, spinors of higher order valence are elements of corresponding univalent spinor *outer products*.

2.2 Complex conjugation of spinor components

A $(p,0;r,0)$ spinor with components $\chi^{A\ldots C}_{D\ldots F}$ belonging to $S^{A\ldots C}_{D\ldots F}$ cannot be considered as the most general kind of spinor, as we have not yet defined the operation of complex conjugation on it. As we have already witnessed in the previous chapter, complex conjugation of spin-vector components arises, otherwise we would not have real-valued world-vector components.

Consider an element ζ^A of S^A. Now, just because S^A is a vector space over the complex field $\mathbb{C}$ does not imply that the complex conjugate of ζ^A is also an element of S^A. In fact, it is not. To see why this assumption is false we need only to form a linear combination $\zeta^A + \overline{\zeta^A}$ (complex conjugation is denoted by a bar above the element), and straight away we would have a real element of S^A. Hence, the complex conjugate of an element ζ^A must belong to an entirely new vector space. To indicate that an element belongs to a conjugated vector space we write $\overline{\zeta} \in \overline{S}$, then the dual is $\overline{[\zeta, \,]} \in \overline{S^*}$. In abstract index notation: given an element $\zeta^A \in S^A$ the complex conjugate of ζ^A is written

$$\zeta^A \mapsto \overline{\zeta^A} = \bar{\zeta}^{A'} \in S^{A'}. \tag{2.1}$$

The prime attached to the abstract index A is indicative that $S^{A'}$ is the complex conjugate of S^A. That $S^{A'}$ is not isomorphic to S^A can be seen by noting that the conjugate of $\alpha\zeta^A + \beta\eta^A$, for $\alpha, \beta \in \mathbb{C}$ and $\zeta^A, \eta^A \in S^A$, is

$$\alpha\zeta^A + \beta\eta^A \mapsto \overline{\alpha\zeta^A + \beta\eta^A} = \bar{\alpha}\bar{\zeta}^{A'} + \bar{\beta}\bar{\eta}^{A'} \tag{2.2}$$

and not $\alpha\bar{\zeta}^{A'} + \beta\bar{\eta}^{A'}$. Then $S^{A'}$ is anti-isomorphic to S^A. Note that a double application of complex conjugation applied to a spinor leaves its original configuration unaltered e.g. $\overline{\bar{\zeta}^{A'}} = \zeta^A$.

An exception to (2.1) is given for the bivalent spinor $\epsilon_{AB} \in S_{AB}$. This is the *Levi–Civita* spinor and will be defined later. In so far as complex conjugation is involved we would conventionally write $\overline{\epsilon_{AB}}$ as $\epsilon_{A'B'} \in S_{A'B'}$ and not $\bar{\epsilon}_{A'B'}$. The omission of the bar above the kernel letter will not cause ambiguities in our notation.

A $(p, q; r, s)$ spinor with components $\chi^{A...CS'...U'}{}_{D...FW'...Y}$ is the most general kind of spinor. It is a bilinear map $\chi^{A...CS'...U'}{}_{D...FW'...Y'} : S_A \times \cdots \times S_C \times S_{S'} \times \cdots \times S_{U'} \times S_D \times \cdots \times S_F \times S_{W'} \times \cdots \times S_{Y'} \to S$ where $\chi^{A...CS'...U'}{}_{D...FW'...Y'} \in S^{A...CS'...U'}{}_{D...FW'...Y'}$ possesses p contravariant unprimed indices, q contravariant primed indices, r covariant unprimed indices and s covariant primed indices. Because S and $\overline{S}$ are not anti-isomorphic to each other and define different vector spaces, the order of primed and unprimed index positions is irrelevant. For example, consider a $(1, 1; 1, 1)$ spinor with components $\kappa^{AB'}{}_{CD'}$, then

$$\kappa^{AB'}{}_{CD'} = \kappa^{B'A}{}_{D'C} = \kappa^{B'}{}_{D'}{}^{A}{}_{C}. \tag{2.3}$$

However, S and S^* are isomorphic to each other and define 'equivalent' vector spaces. The reordering of primed through primed and unprimed through unprimed is forbidden. For example, given a $(2, 0; 2, 0)$ spinor with components $\tau^{AB}{}_{CD}$, then

$$\tau^{AB}{}_{CD} \neq \tau^{BA}{}_{CD} \tag{2.4}$$

nor any other possible combination of reordering. Similarly, for a $(0, 2; 0, 2)$ spinor with $\tau^{A'B'}{}_{C'D'}$ the same rule applies.

2.3 Vector bases and abstract indices

Let $\delta_1^a, \delta_2^a, \ldots \delta_n^a$ constitute a basis of n-elements belonging to the vector space S^a. Then any vector $X^a \in S^a$ can be written as a linear combination of its components $X^1, X^2, \ldots, X^n \in \mathbb{C}$ with respect to this basis:

$$X^a = X^1\delta_1^a + X^2\delta_2^a + \ldots + X^n\delta_n^a, \tag{2.5}$$

which can be expressed in the more compactified form

$$X^a = X^{\mathbf{a}}\delta_{\mathbf{a}}^a. \tag{2.6}$$

Recall that boldface indices are numerical, and range from $1, 2, \ldots, n$ for n-dimensional space. Note that $X^{\mathbf{a}}$ are scalars belonging to $\mathbb{C}$; not a vector belonging to $S^{\mathbf{a}}$. Also, as is true for all dummy indices, $\mathbf{a}$ can be replaced by any other boldface index, e.g. $X^a = X^{\mathbf{i}}\delta_{\mathbf{i}}^a$ etc. The boldface type exhibited in (2.6) indicates two important consequences: (1) summation is presumed, (2) a basis is indicated. These consequences will be adhered to throughout the subsequent chapters.

We can now describe the dual basis of $\delta_{\mathbf{a}}^a$ by $\delta_a^{\mathbf{a}} \in S_a$ where $\delta_a^{\mathbf{a}}$ defines a linear map $\delta_a^{\mathbf{a}} : S^a \to \mathbb{C}$. Then

$$X^{\mathbf{a}} = X^a\delta_a^{\mathbf{a}}. \tag{2.7}$$

Define $\delta_a^{\mathbf{a}}$ to be a basis of n-elements belonging to the vector space S_a, and let any vector $Y_a \in S_a$ be written as a linear combination of its components $Y_1, Y_2, \ldots, Y_n \in \mathbb{C}$ with respect to this basis:

$$Y_a = Y_1\delta_a^1 + Y_2\delta_a^2 + \cdots + Y_n\delta_a^n \tag{2.8}$$

or

$$Y_a = Y_{\mathbf{a}}\delta_a^{\mathbf{a}} \tag{2.9}$$

where

$$Y_{\mathbf{a}} = Y_a\delta_{\mathbf{a}}^a. \tag{2.10}$$

Transvecting (2.6) through by Y_a yields

$$X^aY_a = X^{\mathbf{a}}Y_a\delta_{\mathbf{a}}^a, \tag{2.11}$$

which gives

$$X^aY_a = X^{\mathbf{a}}Y_{\mathbf{a}} \tag{2.12}$$

by (2.10). This suggests that the contracted product $X^a Y_a$ represents a summation of components.

2.4 Levi–Civita spinor

That there exists an element ϵ_{AB} of S_{AB}, called the *Levi–Civita* spinor or *epsilon* spinor, so that

$$[\zeta, \eta] = \epsilon_{AB}\zeta^A \eta^B \tag{2.13}$$

and

$$[\eta, \zeta] = -\epsilon_{AB}\eta^B \zeta^A \tag{2.14}$$

follows from (1.36), (1.37) and (1.38). It is obvious that ϵ_{AB} is skew-symmetrical, i.e.

$$\epsilon_{AB} = -\epsilon_{BA}. \tag{2.15}$$

We will see shortly that ϵ_{AB} can be considered as the spinor counterpart of the metric tensor g_{ab} of standard tensor theory. Indeed, we can treat ϵ_{AB} as a quantity which lowers indices:

$$\zeta_B = \epsilon_{AB}\zeta^A. \tag{2.16}$$

This can be seen from (2.13) or (2.14) — $\epsilon_{AB}\zeta^A$ being the dual of η^B.

We can use (2.13) or (2.14) to obtain the component form of the inner product. This together with (2.16) yields

$$[\zeta, \eta] = \zeta_B \eta^B. \tag{2.17}$$

Although the index B is abstract and therefore acts only as a label, we are permitted to sum if we recall (2.12). Hence,

$$[\zeta, \eta] = \zeta_0 \eta^0 + \zeta_1 \eta^1. \tag{2.18}$$

(Note the summation ranges over $0, 1$ instead of $1, 2$ for the same reasons as those given in Sec. 2.1.) Comparing (1.46) with (2.18) yields the following relations between components of ζ^A and its dual ζ_A:

$$\zeta^0 = \zeta_1, \ \zeta^1 = -\zeta_0. \tag{2.19}$$

These suggest that there exists an element ϵ^{AB} of S^{AB}, such that

$$\zeta^A = \epsilon^{AB}\zeta_B. \tag{2.20}$$

Because of the skew-symmetry property exhibited by the Levi–Civita spinor, ambiguities can arise during calculations unless one pays particular attention to the position of each index. This also applies to the raising and lowering of indices.

The two-dimensional Kronecker symbol can be obtained in a simple way by effectively substituting (2.16) into (2.20) — with the appropriate choice of dummy indices:

$$\epsilon^{AB}\epsilon_{AC} = \delta_C^B = \epsilon_C{}^B \tag{2.21}$$

and similarly,

$$-\epsilon_{BA}\epsilon^{CB} = \delta_A^C = -\epsilon^C{}_A. \tag{2.22}$$

In actuality, one tends to dispense with the conventional Kronecker symbol in preference for $\epsilon_A{}^B$. Note now the specific ordering of the indices on the far right of (2.21) and (2.22). The relation

$$\epsilon_A{}^B = -\epsilon^B{}_A \tag{2.23}$$

is evident, and so too is

$$\epsilon^{AB} = -\epsilon^{BA} \tag{2.24}$$

by (2.15). Thus the rules governing the raising and lowering of spinor indices applied to the components of some multivalent spinor are as follows:

$$\begin{aligned}\chi^{M\ldots NA} &= \epsilon^{AB}\chi^{M\ldots N}{}_B \\ &= -\chi^{M\ldots N}{}_B\epsilon^{BA} \\ &= \chi^{M\ldots NB}\epsilon_B{}^A\end{aligned} \tag{2.25}$$

and,

$$\begin{aligned}\chi^{M\ldots N}{}_A &= -\varepsilon_{AB}\chi^{M\ldots NB} \\ &= \chi^{M\ldots NB}\epsilon_{BA} \\ &= -\chi^{M\ldots N}{}_B\epsilon^B{}_A.\end{aligned} \tag{2.26}$$

If each of the right-hand sides of (2.25) are compared with each of the right-hand sides of (2.26) in turn, there seems to be a simple relationship which yields each in terms of the other. This is achieved through the use of Penrose's 'see-saw':

$$\chi^{M\ldots N}{}_A{}^{AR\ldots S} = -\chi^{M\ldots NA}{}_A{}^{R\ldots S}. \tag{2.27}$$

Whether a minus sign occurs during contraction can be determined by noting the relative positions of the adjacent indices under contraction. If we study (2.25) or (2.26), we notice that when adjacent indices ascend to the left they induce no minus sign in the expression, while adjacent indices which ascend to the right induce a minus sign in the expression.

Of course, everything discussed above applies equally to unprimed as well as primed indices.

2.5 Spinor dyad basis and its components

We have already stated that $\boldsymbol{o}$,$\boldsymbol{\iota}$ constitute a spin basis for S (see (1.42)). Using (2.13) and (2.14) this can be reiterated in terms of the orthonormalisation conditions:

$$[\boldsymbol{o},\boldsymbol{\iota}] = \epsilon_{AB}o^A\iota^B = o_A\iota^A = 1 \tag{2.28}$$

$$[\boldsymbol{\iota},\boldsymbol{o}] = \epsilon_{AB}\iota^B o^A = \iota_A o^A = -1 \tag{2.29}$$

$$[\boldsymbol{o},\boldsymbol{o}] = \epsilon_{AB}o^A o^B = o_A o^A = 0 \tag{2.30}$$

$$[\boldsymbol{\iota},\boldsymbol{\iota}] = \epsilon_{AB}\iota^A\iota^B = \iota_A\iota^A = 0. \tag{2.31}$$

The two basis spinors $o^A, \iota^A \in S^A$ form a *dyad* in spin-space; this is analogous to a basis in Minkowski space forming a tetrad. The dyad can be written collectively as $\epsilon_{\mathbf{A}}{}^A \in S^A$ where

$$\epsilon_0{}^A = o^A, \ \epsilon_1{}^A = \iota^A. \tag{2.32}$$

The product of $\epsilon_{\mathbf{A}}{}^A$ together with the dual basis must be the Kronecker symbol. That is

$$\delta_{\mathbf{A}}{}^{\mathbf{B}} \equiv \epsilon_{\mathbf{A}}{}^{\mathbf{B}} = \epsilon_{\mathbf{A}}{}^A\epsilon_A{}^{\mathbf{B}} = \begin{pmatrix} 1 & 0 \\ 0 & 1 \end{pmatrix}. \tag{2.33}$$

From the orthonormalisation conditions (2.28)–(2.31) the components of ϵ_{AB} with respect to $\epsilon_{\mathbf{A}}{}^A \in S^A$ are

$$\epsilon_{\mathbf{AB}} = \epsilon_{AB}\epsilon_{\mathbf{A}}{}^A\epsilon_{\mathbf{B}}{}^B = \begin{pmatrix} o_A o^A & o_A\iota^A \\ \iota_A o^A & \iota_A\iota^A \end{pmatrix} = \begin{pmatrix} 0 & 1 \\ -1 & 0 \end{pmatrix}. \tag{2.34}$$

Clearly the matrix (2.34) is invertible, and we denote the inverse by

$$-E^{\mathbf{AB}} = \begin{pmatrix} 0 & 1 \\ -1 & 0 \end{pmatrix}. \tag{2.35}$$

We must have

$$-E^{\mathbf{AB}}\epsilon_{\mathbf{AC}} = \epsilon_{\mathbf{C}}{}^{\mathbf{B}} \tag{2.36}$$

which implies from (2.21) that

$$\epsilon^{\mathbf{AB}} = -E^{\mathbf{AB}}. \tag{2.37}$$

So combining (2.34), (2.35) and (2.37) yields

$$\epsilon^{\mathbf{AB}} = \epsilon_{\mathbf{AB}} = \begin{pmatrix} 0 & 1 \\ -1 & 0 \end{pmatrix}, \tag{2.38}$$

which can be expressed as

$$\epsilon^{\mathbf{AB}} = \epsilon^{AB}\epsilon_A{}^{\mathbf{A}}\epsilon_B{}^{\mathbf{B}}. \tag{2.39}$$

If we rewrite (2.33), (2.34) and (2.39) respectively as

$$\epsilon_A{}^B = \epsilon_A{}^{\mathbf{A}}\epsilon_{\mathbf{A}}{}^B \tag{2.40}$$

$$\epsilon_{AB} = \epsilon_{\mathbf{AB}}\epsilon_A{}^{\mathbf{A}}\epsilon_B{}^{\mathbf{B}} \tag{2.41}$$

$$\epsilon^{AB} = \epsilon^{\mathbf{AB}}\epsilon_{\mathbf{A}}{}^A\epsilon_{\mathbf{B}}{}^B \tag{2.42}$$

then it is a straightforward exercise to show that for a dyad basis:

$$\epsilon_A{}^B = o_A\iota^B - \iota_A o^B \tag{2.43}$$

$$\epsilon_{AB} = o_A\iota_B - \iota_A o_B \tag{2.44}$$

$$\epsilon^{AB} = o^A\iota^B - \iota^A o^B. \tag{2.45}$$

For $\zeta^A \in S^A$ we write equation (1.43) as

$$\zeta^A = \zeta^{\mathbf{A}}\epsilon_{\mathbf{A}}{}^A = \zeta^0 o^A + \zeta^1\iota^A \tag{2.46}$$

where o^A and ι^A are given by (2.32). Upon transvection with o_A and ι_A (2.46) yields

$$\zeta^0 = -\iota_A\zeta^A, \ \zeta^1 = o_A\zeta^A. \tag{2.47}$$

Comparing (2.33) with the orthonormalisation conditions (2.28)–(2.31) suggests that

$$\epsilon_A{}^0 = -\iota_A, \ \epsilon_A{}^1 = o_A, \tag{2.48}$$

which is in agreement with (2.47). The components of $\zeta_A \in S_A$ are given by the transvection of

$$\zeta_A = \zeta_{\mathbf{A}} \epsilon_A{}^{\mathbf{A}} \tag{2.49}$$

with o^A and ι^A. Hence

$$\zeta_0 = \zeta_A o^A, \; \zeta_1 = \zeta_A \iota^A. \tag{2.50}$$

By comparing (2.47) with (2.50) we see that

$$\zeta_0 = -\zeta^1, \; \zeta_1 = \zeta^0. \tag{2.51}$$

The above analysis holds if we replace all the quantities by their complex conjugates. In particular we define the complex conjugate of the basis elements $\epsilon_{\mathbf{A}}{}^A \in S^A$ as $\epsilon_{\mathbf{A}'}{}^{A'} \in S^{A'}$, or explicitly as conjugal spinor dyad components:

$$o^{A'} = \bar{o}^{A'} = \overline{o^A}, \; \iota^{A'} = \bar{\iota}^{A'} = \overline{\iota^A} \tag{2.52}$$

where

$$\epsilon_{0'}{}^{A'} = o^{A'}, \; \epsilon_{1'}{}^{A'} = \iota^{A'}. \tag{2.53}$$

Note the absence of the bars in (2.53). Again, as in the case of $\epsilon_{A'B'}$ etc., there will be no ambiguity in our notation.

2.6 Spinor symmetry operations

The identity

$$\epsilon_{A[B}\epsilon_{CD]} = 0 \tag{2.54}$$

where by convention square brackets [ ] denote skew-symmetrisation, plays an important role in two-component spinor algebra. Its use heralded the statement by Penrose, *'only symmetric spinors matter'*.

Let us rewrite (2.54) such that

$$\epsilon_{AB}\epsilon_{CD} + \epsilon_{AC}\epsilon_{DB} + \epsilon_{AD}\epsilon_{BC} = 0. \tag{2.55}$$

Raising the indices C and D yields, after some slight rearranging,

$$\epsilon_A{}^C\epsilon_B{}^D - \epsilon_B{}^C\epsilon_A{}^D = \epsilon_{AB}\epsilon^{CD}. \tag{2.56}$$

If we transvect (contract) with a multivalent spinor $\chi_{\cdots CD \cdots}$ then

$$2\chi_{\cdots[AB]} = \chi_{\cdots AB\cdots} - \chi_{\cdots BA\cdots} = \chi_{\cdots C}{}^{C}{}_{\cdots}\epsilon_{AB}. \tag{2.57}$$

If $\chi_{\cdots AB} = \chi_{\cdots [AB]} \cdots$ then (2.57) becomes

$$2\chi_{\cdots AB \cdots} = \chi_{\cdots C}{}^{C}{}_{\cdots}\epsilon_{AB}. \tag{2.58}$$

Now because for some multivalent spinor

$$\chi_{\cdots AB \cdots} = \chi_{\cdots (AB) \cdots} + \chi_{\cdots [AB] \cdots}, \tag{2.59}$$

where round brackets ( ) denote symmetrisation, it follows from (2.58) that

$$\chi_{\cdots AB \cdots} = \chi_{\cdots (AB) \cdots} + \frac{1}{2}\chi_{\cdots C}{}^{C}{}_{\cdots}\epsilon_{AB}. \tag{2.60}$$

In fact, as will be demonstrated below, any multivalent spinor $\chi_{A \ldots CR' \ldots T'}$ can be composed of a sum incorporating a symmetric part $\chi_{(A \ldots C)(R' \ldots T')}$ and an outer product involving the Levi–Civita spinor and spinors of lower valence. Thus we concur with Penrose's statement above in the sense of (2.60).

If the difference between two equivalent spinors produces a sum consisting of outer products of Levi–Civita spinors and the original spinor but with lower valence, then an equivalence relation exists between the two original spinors. For example, (2.60) suggests that an equivalence relation exists between $\chi_{\cdots AB \cdots}$ and $\chi_{\cdots (AB) \cdots}$ for which we write

$$\chi_{AB} \equiv \chi_{(AB)} (\mathrm{mod} R_2) \tag{2.61}$$

where R_2 is an equivalence relation defined above, but for the bivalent spinor χ_{AB}. We shall show by the method of induction that given a n-valent spinor $\psi_{AB \cdots Z}$

$$\psi_{AB \cdots Z} \equiv \psi_{(AB \ldots Z)} (\mathrm{mod} R_n). \tag{2.62}$$

It is assumed then that given a $(n-1)$-valent spinor $\psi_{B \ldots Z}$

$$\psi_{B \ldots Z} \equiv \psi_{(B \ldots Z)} (\mathrm{mod} R_{n-1}). \tag{2.63}$$

Clearly, for any totally symmetric n-valent spinor one can write

$$\psi_{(ABC \ldots Z)} = \frac{1}{n}[\psi_{A(BC \ldots Z)} + \psi_{B(AC \ldots Z)} + \psi_{C(AB \ldots Z)} + \ldots + \psi_{Z(ABC \ldots Y)}] \tag{2.64}$$

where, of course, there must be n-terms on the right. Let us now take the difference between the first term and the second term on the right of

(2.64), (note we could have taken the difference between any two terms on the right) then from (2.57)

$$\psi_{A(BC...Z)} - \psi_{B(AC...Z)} = -\psi^{X}{}_{(XCD...Z)}\epsilon_{AB}. \tag{2.65}$$

Repeating this process for each term on the right eventually yields

$$\psi_{(ABC...Z)} \equiv \psi_{A(BC...Z)}(\text{mod}R_n). \tag{2.66}$$

Transvecting through (2.66) by an arbitrary univalent spinor ζ^A gives

$$\zeta^A\psi_{ABC...Z} \equiv \zeta^A\psi_{A(BC...Z)}(\text{mod}R_{n-1}) \tag{2.67}$$

by (2.63). Hence,

$$\psi_{ABC...Z} \equiv \psi_{A(BC...Z)}(\text{mod}R_n). \tag{2.68}$$

Finally, comparing this with (2.66) yields (2.62). If we carry out the above procedures using primed indices then one can establish the relation $\psi_{A'B'...Z'} \equiv \psi_{(A'B'...Z')}(\text{mod}R_n)$. Thus the two procedures in combination will result in the statement made earlier.

2.7 The connection between world-tensors and spinors

As we shall discover later, many of general relativity's tensor equations have a very simple and elegant spinor counterpart. Also, the complications inherent in world-tensor manipulation are usually markedly reduced when one translates them into spinor form. However, 2-spinors occur in the spinor representation of $SL(2,\mathbb{C})$, whereas we associate world-tensors with the Lorentz group $L(4)$; hence, the groups are non-isomorphic to each other. Because a homomorphism exists between $SL(2,\mathbb{C})$ and $L(4)$ and not an isomorphism, not all spinors have a tensor counterpart.

2.7.1 *Infeld–van der Waerden symbols*

The connecting quantities which allow transference between world-tensors and spinors are referred to as *Infeld–van der Waerden symbols*, $\sigma^{\mathbf{a}}{}_{AB'}$. Collectively, this symbol represents four 2×2 Hermitian matrices i.e. the three Pauli spin matrices and the unit matrix apart from the factor $2^{-\frac{1}{2}}$:

$$\sigma^0{}_{AB'} = \frac{1}{\sqrt{2}}\begin{pmatrix} 1 & 0 \\ 0 & 1 \end{pmatrix}, \quad \sigma^1{}_{AB'} = \frac{1}{\sqrt{2}}\begin{pmatrix} 0 & 1 \\ 1 & 0 \end{pmatrix}$$

$$\sigma^2{}_{AB'} = \frac{1}{\sqrt{2}}\begin{pmatrix} 0 & i \\ -i & 0 \end{pmatrix}, \quad \sigma^3{}_{AB'} = \frac{1}{\sqrt{2}}\begin{pmatrix} 1 & 0 \\ 0 & -1 \end{pmatrix}. \tag{2.69}$$

Note that $\sigma^{\mathbf{a}}{}_{AB'}$ transform as world-vectors on the index **a**, while on the indices A and B' they transform as spinors. Stating that $\sigma^{\mathbf{a}}{}_{AB'}$ is Hermitian means

$$\overline{\sigma^{\mathbf{a}}{}_{AB'}} = \sigma^{\mathbf{a}}{}_{BA'} \tag{2.70}$$

where, by definition,

$$\overline{\sigma^{\mathbf{a}}{}_{AB'}} = \overline{\sigma}^{\mathbf{a}}{}_{A'B}. \tag{2.71}$$

Note that for a general spinor $\chi_{AA'\ldots}{}^{BB'\ldots}$ to be 'real' it must satisfy the condition

$$\overline{\chi}_{AA'\ldots}{}^{BB'\ldots} = \chi_{AA'\ldots}{}^{BB'\ldots}. \tag{2.72}$$

Clearly for spinors whose unprimed valence is less or greater than its primed valence, the above operation would be nonsensical.

The relationship between the Infeld–van der Waerden symbols and the metric tensor $g_{\mathbf{ab}}$ is defined to be

$$g_{\mathbf{ab}} = \epsilon_{AB}\epsilon_{A'B'}\sigma_{\mathbf{a}}{}^{AA'}\sigma_{\mathbf{b}}{}^{BB'}. \tag{2.73}$$

By introducing the equations

$$\sigma_{\mathbf{a}}{}^{AA'}\sigma^{\mathbf{b}}{}_{AA'} = \delta^{\mathbf{b}}_{\mathbf{a}} \tag{2.74}$$

and

$$\sigma^{\mathbf{a}}{}_{AA'}\sigma_{\mathbf{a}}{}^{BB'} = \epsilon_A{}^B\epsilon_{A'}{}^{B'} \tag{2.75}$$

one can show that (2.73) is equivalent to

$$g_{\mathbf{ab}}\sigma^{\mathbf{a}}{}_{AA'}\sigma^{\mathbf{b}}{}_{BB'} = \epsilon_{AB}\epsilon_{A'B'}. \tag{2.76}$$

If $T_{\mathbf{ab}}{}^{\mathbf{cd}}$ is any world-tensor then the spinor equivalent is defined to be

$$T_{AA'BB'}{}^{CC'DD'} = T_{\mathbf{ab}}{}^{\mathbf{cd}}\sigma^{\mathbf{a}}{}_{AA'}\sigma^{\mathbf{b}}{}_{BB'}\sigma_{\mathbf{c}}{}^{CC'}\sigma_{\mathbf{d}}{}^{DD'}. \tag{2.77}$$

Conversely, the world-tensor counterpart of (2.77) is

$$T_{\mathbf{ab}}{}^{\mathbf{cd}} = T_{AA'BB'}{}^{CC'DD'}\sigma_{\mathbf{a}}{}^{AA'}\sigma_{\mathbf{b}}{}^{BB'}\sigma^{\mathbf{c}}{}_{CC'}\sigma^{\mathbf{d}}{}_{DD'} \tag{2.78}$$

which is obtained from (2.77) by using (2.74).

The spinor equivalent of the metric tensor $g_{\mathbf{ab}}$ is given by (2.76). That is

$$g_{AA'BB'} = g_{\mathbf{ab}}\sigma^{\mathbf{a}}{}_{AA'}\sigma^{\mathbf{b}}{}_{BB'} \tag{2.79}$$

and

$$g^{AA'BB'} = g^{\mathbf{ab}}\sigma_{\mathbf{a}}{}^{AA'}\sigma_{\mathbf{b}}{}^{BB'}. \tag{2.80}$$

Recall (see (2.3)) that spinor indices can be reordered provided the reordering takes place between primed and unprimed indices. Thence, from (2.79) and (2.80)

$$g_{ABA'B'} = \epsilon_{AB}\varepsilon_{A'B'},\ g^{ABA'B'} = \epsilon^{AB}\epsilon^{A'B'}. \tag{2.81}$$

The Infeld–van der Waerden symbols, although they provide the means to 'transform' world-tensors into their spinor equivalent and vice-versa, are seldom used in general relativistic calculations explicitly or otherwise. Indeed, it is conventional to dispense with the symbols entirely and write, for example, the metric tensor as

$$g_{ab} = \epsilon_{AB}\epsilon_{A'B'} \tag{2.82}$$

$$g^{ab} = \epsilon^{AB}\epsilon^{A'B'} \tag{2.83}$$

and the Kronecker delta as

$$\delta_a^b = \epsilon_A{}^B\epsilon_{A'}{}^{B'}. \tag{2.84}$$

(Note that in some older texts the tensor-spinor correspondence utilises '$\leftrightarrow$' in place of '='. The former convention has been largely superseded by the latter.)

2.7.2 *Null vectors*

Let Z^a be any world-vector in space-time. Its spinor equivalent is

$$Z^a = Z^{AB'}. \tag{2.85}$$

Z_a is null, i.e.

$$Z_aZ^a = 0, \tag{2.86}$$

if and only if $Z^{AB'}$ can be expressed in terms of a product of two univalent spinors, one unprimed, the other primed:

$$Z^{AB'} = \zeta^A \eta^{B'}. \tag{2.87}$$

If Z^a is real, i.e.

$$Z^a = \overline{Z}^a, \tag{2.88}$$

then $Z^{AB'}$ is Hermitian

$$Z^{AB'} = \overline{Z}^{B'A}. \tag{2.89}$$

The reality of Z^a implies, from (2.87), that $\eta^{B'} = \lambda\overline{\zeta}^{B'}$ for $\lambda \in \mathbb{R}$, provided $\overline{\zeta}^{B'} \neq 0$. Thus

$$Z^{AB'} = \lambda \zeta^A \overline{\zeta}^{B'} \tag{2.90}$$

from (2.87), or

$$\zeta^A \eta^{B'} = \overline{\zeta}^{B'} \overline{\eta}^A \tag{2.91}$$

from (2.89). If we absorb the factor $|\lambda|^{\frac{1}{2}}$ into ζ^A then (2.90) becomes

$$Z^{AB'} = \pm \zeta^A \overline{\zeta}^{B'}. \tag{2.92}$$

Indeed, every real null world-vector can be expressed in one of the forms

$$Z^a = \pm \zeta^A \overline{\zeta}^{B'}. \tag{2.93}$$

This implies that Z^a is future-pointing for positive (2.93) and past-pointing for the negative value. Consequently, the transition from spinors at a point to spinor fields requires space-time to be time-orientated.

Let us consider how the standard null vectors of general relativity relate to some of the analysis above. This will be of importance when discussion of *spin coefficients* arises in Chapter 3.

Let us introduce a basis $(\boldsymbol{o},\boldsymbol{\iota})$ in spin-space satisfying the normalisation condition

$$o_A \iota^A = 1. \tag{2.94}$$

The null tetrad induced by $(\boldsymbol{o},\boldsymbol{\iota})$ is defined by

$$l^a = o^A o^{A'},\ l_a = o_A o_{A'}$$
$$n^a = \iota^A \iota^{A'},\ n_a = \iota_A \iota_{A'}$$

$$m^a = o^A \iota^{A'}, \; m_a = o_A \iota_{A'}$$
$$\overline{m}^a = \iota^A o^{A'}, \; \overline{m}_a = \iota_A o_{A'} \tag{2.95}$$

where l^a, n^a, m^a and $\overline{m}^a$ are null vectors that are basis forming. They satisfy the following orthonormalisation conditions

$$l^a n_a = -m^a \overline{m}_a = 1$$
$$l^a l_a = n^a n_a = m^a m_a = \overline{m}^a \overline{m}_a = 0$$
$$l^a m_a = l^a \overline{m}_a = n^a m_a = n^a \overline{m}_a = 0. \tag{2.96}$$

Note that l^a and n^a are real whereas m^a and $\overline{m}^a$ are complex conjugates. It is straightforward to show, with the aid of (2.43), (2.44) and (2.45) that

$$g_a{}^b = n_a l^b + l_a n^b - \overline{m}_a m^b - m_a \overline{m}^b \tag{2.97}$$

$$g_{ab} = n_a l_b + l_a n_b - \overline{m}_a m_b - m_a \overline{m}_b \tag{2.98}$$

$$g^{ab} = n^a l^b + l^a n^b - \overline{m}^a m^b - m^a \overline{m}^b. \tag{2.99}$$

Let us introduce at every point in space-time a set of four mutually orthonormal world-vectors $\lambda_{(m)}{}^a$ that are basis forming. The bracketed index represents the tetrad index which numbers the vectors $0, 1, 2, 3$. If we identify with this tetrad the basis vectors of a local Minkowski coordinate system, i.e.

$$g_{(m)(n)} = \lambda_{(m)}{}^a \lambda_{(n)}{}^b g_{ab} = \begin{pmatrix} 1 & 0 & 0 & 0 \\ 0 & -1 & 0 & 0 \\ 0 & 0 & -1 & 0 \\ 0 & 0 & 0 & -1 \end{pmatrix} \tag{2.100}$$

then the orthonormal tetrad will consist of one timelike vector $\lambda_{(0)}{}^a$ and three spacelike vectors $\lambda_{(i)}{}^a (i = 1, 2, 3)$, which are labelled t^a, x^a, y^a and z^a respectively. The inverse of (2.100) is defined to be

$$g_{(n)(p)} g^{(m)(p)} = \delta^{(m)}_{(n)}, \tag{2.101}$$

and together with (2.100), we have

$$g_{ab} = g_{(m)(n)} \lambda^{(m)}{}_a \lambda^{(n)}{}_b. \tag{2.102}$$

From (2.100) and (2.102) we can write the metric tensor g_{ab} as

$$g_{ab} = t_a t_b - x_a x_b - y_a y_b - z_a z_b, \tag{2.103}$$

and it is not difficult to show that

$$g^{ab} = t^a t^b - x^a x^b - y^a y^b - z^a z^b \tag{2.104}$$

or

$$g_a{}^b = t_a t^b - x_a x^b - y_a y^b - z_a z^b. \tag{2.105}$$

Furthermore, the orthonormalisation conditions are easily obtained from (2.100):

$$t^a t_a = -x^a x_x = -y^a y_a = -z^a z_a = 1 \tag{2.106}$$

and

$$t^a x_a = t^a y_a = t^a z_a = x^a y_a = y^a z_a = z^a x_a = 0. \tag{2.107}$$

The relationships between the null tetrad and the orthonormal tetrad are given by

$$\begin{aligned} l^a &= \frac{1}{\sqrt{2}}(t^a + z^a) \\ n^a &= \frac{1}{\sqrt{2}}(t^a - z^a) \\ m^a &= \frac{1}{\sqrt{2}}(x^a - iy^a) \\ \overline{m}^a &= \frac{1}{\sqrt{2}}(x^a + iy^a) \end{aligned} \tag{2.108}$$

and

$$\begin{aligned} t^a &= \frac{1}{\sqrt{2}}(l^a + n^a) \\ x^a &= \frac{1}{\sqrt{2}}(m^a + \overline{m}^a) \\ y^a &= \frac{i}{\sqrt{2}}(m^a - \overline{m}^a) \\ z^a &= \frac{1}{\sqrt{2}}(l^a - n^a). \end{aligned} \tag{2.109}$$

2.8 The decomposition of spinors

We have seen thus far that via the Infeld–van der Waerden symbols one can write down the spinor equivalent of any world-tensor. Indeed, the Infeld–van der Waerden symbols are rarely used in explicit calculations so any tensor can be directly related to its spinor equivalent. One important feature of our discussion which has not yet been addressed, however, is that of the symmetry properties which certain world-tensors, and therefore, also their spinor equivalents, possess.

2.8.1 *The spinor equivalent of a symmetric valence two tensor*

The spinor equivalent of a symmetric covariant world-tensor of valence two which may be complex,

$$T_{ab} = T_{ba}, \tag{2.110}$$

is given by

$$T_{AA'BB'} = T_{BB'AA'}, \tag{2.111}$$

or

$$T_{AA'BB'} = \frac{1}{2}(T_{AA'BB'} + T_{BB'AA'}). \tag{2.112}$$

Because there is no meaning to be attached to the relative order between primed and unprimed indices, (2.112) can be rewritten in the form

$$T_{ABA'B'} = T_{BAB'A'} = \frac{1}{2}(T_{ABA'B'} + T_{BAB'A'}) \tag{2.113}$$

or,

$$T_{ABA'B'} = \frac{1}{2}(T_{ABA'B'} + T_{ABB'A'}) + \frac{1}{2}(T_{BAB'A'} - T_{ABB'A'}). \tag{2.114}$$

As the first parenthesis is symmetric in A', B' and the second parenthesis is skew-symmetric in A, B, (2.114) can be rewritten as

$$T_{ABA'B'} = T_{AB(A'B')} + T_{[BA]B'A'} \tag{2.115}$$

for which we can write

$$T_{ABA'B'} = T_{(AB)(A'B')} + T_{[BA][B'A']}. \tag{2.116}$$

This last expression is true because of the symmetry exhibited in (2.111). A double application of (2.59) and (2.60) to $T_{[BA][B'A']}$ in (2.116) yields

$$T_{ABA'B'} = P_{ABA'B'} + \frac{1}{4}\epsilon_{AB}\epsilon_{A'B'}T_C{}^C{}_{C'}{}^{C'} \tag{2.117}$$

where

$$P_{ABA'B'} = T_{(AB)(A'B')}. \tag{2.118}$$

The spinor $P_{ABA'B'}$ is Hermitian and corresponds to the trace-free part of T_{ab}, i.e. P_{ab}. The second term on the right of (2.117) is real and proportional to the metric tensor and corresponds to the trace of T_{ab} i.e. $\frac{1}{4}g_{ab}T_c{}^c$.

2.8.2 *The spinor equivalent of the electromagnetic field tensor*

As an example of the decomposition of a skew-symmetric world-tensor of valence two, or bivector, we will consider in particular the *electromagnetic field tensor* F_{ab} which is real. Choosing this particular world-tensor instead of any arbitrary one will allow us the opportunity to introduce the spinor classification scheme according to the coincidences between *principal null directions* — this term will be qualified presently.

As the electromagnetic field tensor F_{ab} has symmetry

$$F_{ab} = -F_{ba} \tag{2.119}$$

its spinor equivalent is given by

$$F_{AA'BB'} = -F_{BB'AA'}. \tag{2.120}$$

Carrying out a similar analysis which yielded (2.114), we find that

$$F_{ab} = F_{ABA'B'} = \frac{1}{2}(F_{ABA'B'} - F_{ABB'A'}) + \frac{1}{2}(F_{ABB'A'} - F_{BAB'A'}). \tag{2.121}$$

The first parenthesis is skew-symmetric in A', B', while the second parenthesis is skew-symmetric in A, B. Thus (2.121) can be rewritten as

$$F_{ab} = F_{ABA'B'} = F_{AB[A'B']} + F_{[AB]B'A'}, \tag{2.122}$$

or, on employing (2.57),

$$F_{ab} = F_{ABA'B'} = \frac{1}{2}\epsilon_{A'B'}F_{ABC'}{}^{C'} + \frac{1}{2}\epsilon_{AB}F_C{}^C{}_{B'A'}. \tag{2.123}$$

Clearly, $F_{ABC'}{}^{C'}$ and $F_C{}^C{}_{B'A'}$ must be symmetric in A, B and B', A' respectively from (2.120). Hence, we define an arbitrary spinor ϕ_{AB} to be symmetric such that

$$\phi_{AB} = \phi_{(AB)} = \frac{1}{2}F_{ABC'}{}^{C'}, \tag{2.124}$$

then taking the complex conjugate of this yields

$$\overline{\phi}_{A'B'} = \overline{\phi}_{(A'B')} = \frac{1}{2}\overline{F}_{A'B'C}{}^{C} = \frac{1}{2}F_{C}{}^{C}{}_{A'B'}. \tag{2.125}$$

The last equality follows because $F_{ABA'B'}$ is necessarily Hermitian. Thus (2.123) can now be written, with the aid of (2.124) and (2.125), in the form

$$F_{ab} = F_{ABA'B'} = \phi_{AB}\epsilon_{A'B'} + \epsilon_{AB}\overline{\phi}_{A'B'} \tag{2.126}$$

where ϕ_{AB} is the *electromagnetic spinor* or *Maxwell spinor.* Notice that from the decomposition which resulted in (2.126), instead of needing to consider six real components of F_{ab}, we have only to consider three complex components of ϕ_{AB} when performing calculations, i.e.

$$\phi_0 = \phi_{00},\ \phi_1 = \phi_{01} = \phi_{10},\ \phi_2 = \phi_{11}. \tag{2.127}$$

Before we continue with our analysis, it will be useful if we return, briefly, to the Levi–Civita spinor. The Levi–Civita tensor has a spinor equivalent given by

$$\epsilon_{abcd} = \epsilon_{AA'BB'CC'DD'} = i(\epsilon_{AC}\epsilon_{BD}\epsilon_{A'D'}\epsilon_{B'C'} - \epsilon_{AD}\epsilon_{BC}\epsilon_{A'C'}\epsilon_{B'D'}). \tag{2.128}$$

This is obtained by carrying out a similar analysis which yielded (2.117) and (2.126), and will be left as an exercise for the reader. Raising c, d and hence C, C', D, D' gives

$$\epsilon_{ab}{}^{cd} = i(\epsilon_A{}^C\epsilon_B{}^D\epsilon_{A'}{}^{D'}\epsilon_{B'}{}^{C'} - \epsilon_A{}^D\epsilon_B{}^C\epsilon_{A'}{}^{C'}\epsilon_{B'}{}^{D'}). \tag{2.129}$$

Now let us define the left-dual of F_{ab} by

$${}^*F_{ab} = \frac{1}{2}\epsilon_{ab}{}^{cd}F_{cd}, \tag{2.130}$$

then with the aid of (2.126) and (2.129) we have

$${}^*F_{ABA'B'} = -i\phi_{AB}\epsilon_{A'B'} + i\epsilon_{AB}\overline{\phi}_{A'B'} \tag{2.131}$$

which is the spinor equivalent of (2.130).

We can further decompose (2.126) into parts which are termed *anti-self-dual* and *self-dual.* For if F_{ab} is anti-self-dual, i.e.

$${}^*F_{ab} = -iF_{ab}, \tag{2.132}$$

then the spinor equivalent is

$$F_{ABA'B'} = -F_{ABB'A'}. \tag{2.133}$$

Conversely, if F_{ab} is self-dual, i.e.

$$^*F_{ab} = iF_{ab}, \tag{2.134}$$

then the spinor equivalent can be written as

$$F_{ABA'B'} = F_{ABB'A'}. \tag{2.135}$$

Equations (2.126) and (2.130) imply that

$$\frac{1}{2}(F_{ab} + i^*F_{ab}) = \phi_{AB}\epsilon_{A'B'} \tag{2.136}$$

and

$$\frac{1}{2}(F_{ab} - i^*F_{ab}) = \epsilon_{AB}\overline{\phi}_{A'B'}. \tag{2.137}$$

The first and second terms on the right of (2.123) represent the symmetry properties exhibited in (2.133) and (2.134) respectively. Thence, comparing (2.136) with (2.133), and (2.127) with (2.135), we can define a pair of bivectors A_{ab} and S_{ab} such that

$$A_{ab} = \phi_{AB}\epsilon_{A'B'} \tag{2.138}$$

represents the anti-self-dual part of F_{ab}, and

$$S_{ab} = \epsilon_{AB}\overline{\phi}_{A'B'} \tag{2.139}$$

represents the self-dual part. Evidently, $A_{ab} = \overline{S}_{ab}$. Furthermore, F_{ab} can be written as the sum of these two bivectors:

$$F_{ab} = A_{ab} + S_{ab}. \tag{2.140}$$

2.9 The canonical decomposition of symmetric spinors

In this section it will be shown how to classify the electromagnetic spinor and the Weyl spinor (yet to be defined) according to the multiplicities of principal null directions. In order for this to be achieved, it will be necessary to show that a symmetric spinor can be decomposed into a symmetrised product of univalent spinors. However, it will be sufficient for our purposes to delineate two specific examples of a more general result.

2.9.1 Classification of the electromagnetic spinor

We wish to decompose the electromagnetic spinor ϕ_{AB} into a symmetrised product of two univalent spinors.

Let $\xi^A \in S^A$ be an arbitrary spinor with components relative to some spinor basis, ξ^0 and ξ^1. Now, the expression $\phi_{AB}\xi^A\xi^B$ is a second degree homogeneous polynomial in ξ^0 and ξ^1, which can be factorised into two linear factors since we are confined in the complex field, which is algebraically closed. Thus

$$\begin{aligned}\phi_{AB}\xi^A\xi^B &= \phi_{00}\xi^0\xi^0 + 2\phi_{10}\xi^1\xi^0 + \phi_{11}\xi^1\xi^1 \\ &= (\xi^1)^2[\phi_{00}K^2 + 2\phi_{10}K + \phi_{11}] \\ &= (\xi^1)^2(\alpha_0 K + \alpha_1)(\beta_0 K + \beta_1) \\ &= (\alpha_A\xi^A)(\beta_B\xi^B) \end{aligned} \tag{2.141}$$

where $K = \xi^0/\xi^1$. Because ξ^A is arbitrary it follows that

$$\phi_{AB} = \alpha_{(A}\beta_{B)} \tag{2.142}$$

is called the *canonical decomposition* of ϕ_{AB}. The spinors α_A and β_B are called *principal spinors*. Each of them can determine a real null direction called a *principal null direction*, (there are at most two of these). We classify the electromagnetic spinor according to its *type*. If the principal spinors are proportional to each other then ϕ_{AB} is termed type N or *algebraically special*; otherwise, we call it type I or *algebraically general*.

Type	Partition	$\phi_{AB} =$
I	$\{11\}$	$\alpha_{(A}\beta_{B)}$
N	$\{2\}$	$\alpha_A\alpha_B$
O	$\{-\}$	0

Table 2.1 Classification of ϕ_{AB} by principal null directions

(Note that the *partition* indicates the two possible coincidences for the principal null directions. The last row is included for completeness.)

2.9.2 Classification of the Weyl spinor

The classification of the Weyl curvature tensor is more commonly known as the *Petrov classification*. Consider the eigenvalue equation

$$F_{ab}V^b = \lambda V_a \tag{2.143}$$

where F_{ab} is the electromagnetic field tensor and V_a are corresponding eigenvectors. It is readily seen, by contracting with V^a, that eigenvectors corresponding to non-zero eigenvalues must be null. It follows from

$$V_{[c}F_{a]b}V^b = 0 \tag{2.144}$$

that F_{ab} may now be classified according to the multiplicities of the null eigenvectors V_a. Analogously, consider the equation

$$V_{[e}C_{a]bc[d}V_{f]}V^bV^c = 0 \tag{2.145}$$

where C_{abcd} is the Weyl conformal tensor with symmetry properties

$$C_{abcd} = C_{[ab][cd]} = C_{cdab}. \tag{2.146}$$

There exist, at most, four null directions or null eigenvectors $V_a \neq 0$ for C_{abcd} which satisfy (2.145). C_{abcd} can be classified according to the multiplicities of V_a. Table 2.2 illustrates the Petrov classification according to the multiplicities of the principal null directions.

Petrov type	Multiplicities	Condition for V_a being a P.N.D.
I	$(1,1,1,1)$	$V_{[e}C_{a]bc[d}V_{f]}V^bV^c = 0$
II	$(2,1,1)$	$V_{[e}C_{a]bcd}V^bV^c = 0$
D	$(2,2)$	$V_{[e}C_{a]bcd}V^bV^c = 0$
III	$(3,1)$	$V_{[e}C_{a]bdc}V^b = 0$
N	(4)	$C_{abcd}V^a=0$
O	–	$C_{abcd} = 0$

Table 2.2 Petrov classification according to the multiplicities of principal null directions

The spinor classification of the Weyl tensor is a more elegant method than the preceding approach, and is analogous to the spinor classification of the electromagnetic field tensor. Let $\xi^A \in S^A$ be an arbitrary spinor with components, on choosing an arbitrary spinor basis, ξ^0 and ξ^1. Now consider the fourth degree homogeneous polynomial in ξ^0 and ξ^1,

$$\begin{aligned}\Psi_{ABCD}\xi^A\xi^B\xi^C\xi^D &= \Psi_{0000}\xi^0\xi^0\xi^0\xi^0 + 4\Psi_{1000}\xi^1\xi^0\xi^0\xi^0 \\ &\quad +6\Psi_{1100}\xi^1\xi^1\xi^0\xi^0 + 4\Psi_{0111}\xi^0\xi^1\xi^1\xi^1 + \Psi_{1111}\xi^1\xi^1\xi^1\xi^1 \\ &= (\xi^1)^4[\Psi_{0000}K^4 + 4\Psi_{1000}K^3 \\ &\quad +6\Psi_{1100}K^2 + 4\Psi_{0111}K + \Psi_{1111}],\end{aligned} \tag{2.147}$$

where Ψ_{ABCD} is the Weyl spinor — yet to be defined. Factorising yields

$$\begin{aligned}\Psi_{ABCD}\xi^A\xi^B\xi^C\xi^D &= (\xi^1)^4(\alpha_0 K + \alpha_1)(\beta_0 K + \beta_1)(\gamma_0 K + \gamma_1)(\delta_0 K + \delta_1)\\ &= (\alpha_A\xi^A)(\beta_B\xi^B)(\gamma_C\xi^C)(\delta_D\xi^D). \qquad (2.148)\end{aligned}$$

Because ξ^A is arbitrary we have

$$\Psi_{ABCD} = \alpha_{(A}\beta_B\gamma_C\delta_{D)}. \qquad (2.149)$$

Equation (2.149) is the canonical decomposition of Ψ_{ABCD} and the spinors $\alpha_A, \beta_B, \gamma_C$ and δ_D are principal spinors. Each of these spinors can determine a real null direction (principal null direction) of which there are at most four. Table 2.3 delineates the classification.

Petrov type	Partition	$\Psi_{ABCD} =$	Ψ_{ABCD} satisfies:
I	$\{1111\}$	$\alpha_{(A}\beta_B\gamma_C\delta_{D)}$	$\Psi_{ABCD}\xi^A\xi^B\xi^C\xi^D = 0$
II	$\{211\}$	$\alpha_{(A}\alpha_B\gamma_C\delta_{D)}$	$\Psi_{ABCD}\xi^A\xi^B\xi^C = 0$
D	$\{22\}$	$\alpha_{(A}\alpha_B\beta_C\beta_{D)}$	$\Psi_{ABCD}\xi^A\xi^B\xi^C = 0$
III	$\{31\}$	$\alpha_{(A}\alpha_B\alpha_C\beta_{D)}$	$\Psi_{ABCD}\xi^A\xi^B = 0$
N	$\{4\}$	$\alpha_A\alpha_B\alpha_C\alpha_D$	$\Psi_{ABCD}\xi^A = 0$
O	$\{-\}$	0	$\Psi_{ABCD} = 0$

Table 2.3 Classification of the Weyl spinor

In what follows we refer to both Table 2.2 and Table 2.3. The symbols D and N in the 'Petrov type' column are degenerate and null types respectively. It is more common now, however, to call D double due to the multiplicities of the principal null directions. Type N has been directly taken from the electromagnetic field tensor classification to describe the coincidence of all principal null directions. All types except I are algebraically special, type I alone being algebraically general. In all cases, including Table 2.1, type O represents Minkowski space.

It is not difficult to show that the last column in Table 2.2 is equivalent to the last column in Table 2.3. However, we first require the spinor equivalent of the Weyl tensor to achieve this equilibrium, which will be given in section (3.23).

Glancing down through the Petrov types, in either Table 2.2 or Table 2.3, one sees that the degeneracy increases. This degeneracy corresponds to increasing specialisation of the Petrov types which, due to Penrose, can be represented by the following diagram.

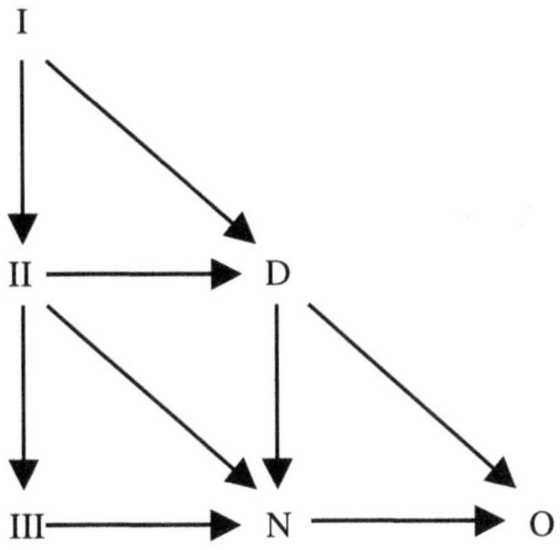

Fig. 2.1 Penrose diagram — hierarchy of Petrov types

2.10 Exercises

2.1 Derive equations (2.43), (2.44) and (2.45).

2.2 Prove that for a spin basis

$$\epsilon_{0A} = o_A = -\epsilon_{A0}$$
$$\epsilon^{0A} = \iota_A = -\epsilon^{A0}$$
$$\epsilon_{1A} = \iota_A = -\epsilon_{A1}$$
$$\epsilon^{1A} = -o^A = -\epsilon^{A1}.$$

2.3 Check the validity of the identity (2.55).

2.4 Show that $\epsilon_A{}^A = 2 = -\epsilon^A{}_A$.

2.5 Let ζ_A and η_B be arbitrary univalent spinors. Show that

$$\zeta_A = \lambda\eta_A \Leftrightarrow \zeta_A\eta^A = 0, \quad \lambda \in \mathbb{C}.$$

2.6 Show that the arbitrary trivalent spinor θ_{ABC} can be written as the sum of the symmetric spinor $\theta_{(ABC)}$ and the outer products of Levi–Civita spinors with symmetric spinors of lower cadence:

$$\theta_{ABC} = \theta_{(ABC)} - \frac{1}{3}\epsilon_{AD}\theta^X{}_{(XC)} - \frac{1}{3}\epsilon_{AC}\theta^X{}_{(XB)} - \frac{1}{2}\epsilon_{BC}\theta_A{}^X{}_X.$$

2.7 Verify equations (2.108) and (2.109).

2.8 Derive equation (2.128).

2.9 Prove relations (2.133), (2.135), (2.136) and (2.137).

Chapter 3

Spinor analysis

So far, we have concentrated on spinors *at a point* throughout our discussion. It is now time to consider spinor fields and to define the concept of the spinor covariant derivative, which, in turn, will introduce us to a new set of quantities to replace the familiar connexion coefficients.

3.1 Spinor form of the covariant derivative

In the following all the required properties which the *spinor covariant derivative* should possess will be defined in an axiomatic fashion.

Define the spinor covariant derivative

$$\nabla_a = \nabla_{AA'} \tag{3.1}$$

by a map

$$\nabla_a : S^{B\ldots C'}_{D\ldots E'} \to S^{B\ldots C'}_{AA'B\ldots E'} \tag{3.2}$$

or,

$$\nabla_a : \zeta_{D\ldots E'}{}^{B\ldots C'} \mapsto \nabla_{AA'}\zeta_{D\ldots E'}{}^{B\ldots C'} \tag{3.3}$$

where $\zeta_{D\ldots E'}{}^{B\ldots C'} \in S^{B\ldots C'}_{D\ldots E'}$ is a general spinor field. The properties satisfied by (3.2) or (3.3) are:

(1) *linearity*:

$$\nabla_a(\zeta_{D\ldots E'}{}^{B\ldots C'} + \eta_{D\ldots E'}{}^{B\ldots C'}) = \nabla_a\zeta_{D\ldots E'}{}^{B\ldots C'} + \nabla_a\eta_{D\ldots E'}{}^{B\ldots C'}, \tag{3.4}$$

where $\eta_{D\ldots E'}{}^{B\ldots C'} \in S^{B\ldots C'}_{D\ldots E'}$ is another general spinor field.

(2) *Liebniz's law:*

$$\nabla_a(\zeta_{D\ldots E'}{}^{B\ldots C'}\phi_{H\ldots I'}{}^{F\ldots G'})$$

$$= \zeta_{D\ldots E'}{}^{B\ldots C'}\nabla_a\phi_{H\ldots I'}{}^{F\ldots G'} + \phi_{H\ldots I'}{}^{F\ldots G'}\nabla_a\zeta_{D\ldots E'}{}^{B\ldots C'}. \tag{3.5}$$

(3) *Commutation with index substitution.*

(4) *Commutation with contraction:*

$$\nabla_a\zeta_{\ldots X\ldots}{}^{\ldots X\ldots} = \epsilon_X{}^Y\nabla_a\zeta_{\ldots Y\ldots}{}^{\ldots X\ldots} = \nabla_a\epsilon_X{}^Y\zeta_{\ldots Y\ldots}{}^{\ldots X\ldots}, \tag{3.6}$$

and similarly for $\epsilon_{x'}{}^{y'}$.

(5) *Commutation with complex conjugation:*

$$\overline{\nabla_a\zeta_{D\ldots E'}{}^{B\ldots C'}} = \nabla_a\overline{\zeta_{D\ldots E'}{}^{B\ldots C'}}, \tag{3.7}$$

implying reality of the covariant derivative

$$\overline{\nabla}_a = \nabla_a. \tag{3.8}$$

(6) ϵ_{AB} *to be covariantly constant:*

$$\nabla_a\epsilon_{BC} = 0 \text{ and } \nabla_a\epsilon^{BC} = 0. \tag{3.9}$$

(7) *Vanishing torsion:*

$$(\nabla_a\nabla_b - \nabla_b\nabla_a)\lambda = 0 \tag{3.10}$$

for all scalars $\lambda \in \mathbb{C}$.

(8) *Partial derivatives operate on scalars:*

$$\nabla_a\lambda = \partial_a\lambda. \tag{3.11}$$

Note that properties (3) and (4) must not, of course, be performed with quantities possessing indices identical to the covariant derivative itself. It should also be noted that the above eight properties are not completely independent, but rather highlight the important features of the spinor covariant derivative operation. Furthermore, in certain scenarios, properties (6) and (7) need not be adhered to. Indeed, the torsion becomes significant when discussing, for example, the Einstein–Cartan–Sciama–Kibble theory.

(This theory is a modification of Einstein's theory where the torsion of space-time is related to the spin density of its matter content. We will not enter into any further examination of this theory here.)

3.2 The curvature spinors

The *Riemann tensor* R_{abcd} possesses the following symmetry properties:

$$R_{[ab][cd]} = R_{abcd} \tag{3.12}$$

$$R_{cdab} = R_{abcd} \tag{3.13}$$

$$R_{a[bcd]} = 0. \tag{3.14}$$

Writing the spinor equivalent as

$$R_{AA'BB'CC'DD'} = R_{abcd}, \tag{3.15}$$

(which, of course, must exhibit the same symmetry properties as those given in (3.12), (3.13) and (3.14) when appropriate unprimed/primed index pairs are interchanged), we can decompose (3.15) in a way analogous to the decomposition (2.126). Thus, noting in (3.12) the skew-symmetry in a, b, we can write

$$\begin{aligned} R_{abcd} = & \frac{1}{2}(R_{ABA'B'CDC'D'} - R_{ABB'A'CDC'D'}) \\ & + \frac{1}{2}(R_{ABB'A'CDC'D'} - R_{BAB'A'CDC'D'}). \end{aligned} \tag{3.16}$$

The first parenthesis is skew-symmetric in A', B' and the second parenthesis is skew-symmetric in A, B. Hence,

$$R_{abcd} = R_{AB[A'B']CDC'D'} + R_{[AB]B'A'CDC'D'}. \tag{3.17}$$

Now, noting in (3.12) the skew-symmetry in c, d, (3.17) becomes

$$\begin{aligned} R_{abcd} = & \frac{1}{2}(R_{AB[A'B']CDC'D'} - R_{AB[A'B']CDD'C'}) \\ & + \frac{1}{2}(R_{AB[A'B']CDD'C'} - R_{AB[A'B']DCD'C'}) \\ & + \frac{1}{2}(R_{[AB]B'A'CDC'D'} - R_{[AB]B'A'CDD'C'}) \\ & + \frac{1}{2}(R_{[AB]B'A'CDD'C'} - R_{[AB]B'A'DCD'C'}). \end{aligned} \tag{3.18}$$

The first and third parentheses in (3.18) are skew-symmetric in C', D' and the second and fourth parentheses are skew-symmetric in C, D. It follows then that

$$R_{abcd} = R_{AB[A'B']CD[C'D']} + R_{AB[A'B'][CD]C'D'} + R_{[AB]A'B'CD[C'D']} + R_{[AB]A'B'[CD]C'D'}. \tag{3.19}$$

By applying (2.57) twice to (3.19) we get

$$R_{abcd} = X_{ABCD}\epsilon_{A'B'}\epsilon_{C'D'} + \Phi_{ABC'D'}\epsilon_{A'B'}\epsilon_{CD} + \overline{\Phi}_{A'B'CD}\epsilon_{AB}\epsilon_{C'D'} + \overline{X}_{A'B'C'D'}\epsilon_{AB}\epsilon_{CD} \tag{3.20}$$

where,

$$X_{ABCD} = \frac{1}{4}R_{ABE'}{}^{E'}{}_{CDF'}{}^{F'}, \tag{3.21}$$

and

$$\Phi_{ABC'D'} = \frac{1}{4}R_{ABE'}{}^{E'}{}_{F}{}^{F}{}_{C'D'}. \tag{3.22}$$

The two spinors X_{ABCD} and $\Phi_{ABC'D'}$ are called *curvature spinors* and uniquely define the spinor equivalent of the Riemann tensor. The coupling with their conjugates in (3.20) is necessary because the Riemann tensor is real.

3.2.1 *Symmetries of the curvature spinors*

The symmetry properties of X_{ABCD} and $\Phi_{ABC'D'}$ can be established directly from the symmetry properties of the Riemann tensor (3.12), (3.13) and (3.14).

The skew-symmetry in a, b of equation (3.12) yields directly, with the aid of (3.20), the following symmetry properties of the curvature spinors: $X_{ABCD} = X_{BACD}$ and $\Phi_{ABC'D'} = \Phi_{BAC'D'}$. Next, the skew-symmetry in c, d allows us to write: $X_{ABCD} = X_{ABDC}$ and $\Phi_{ABC'D'} = \Phi_{ABD'C'}$. By combining these symmetries together we can write

$$X_{ABCD} = X_{(AB)(CD)} \tag{3.23}$$

and

$$\Phi_{ABC'D'} = \Phi_{(AB)(C'D')}. \tag{3.24}$$

Under the interchange of the first and second pairs of indices given by the equation (3.13) we obtain

$$X_{ABCD} = X_{CDAB} \tag{3.25}$$

and

$$\Phi_{ABC'D'} = \overline{\Phi}_{ABC'D'}. \tag{3.26}$$

(Note that (3.26) would actually read $\Phi_{ABC'D'} = \overline{\Phi}_{C'D'AB}$, but because reordering of primed indices through unprimed indices is a legitimate operation (see (2.3)), then (3.26) follows.)

Finally, we wish to establish that the cyclic symmetry exhibited by the Riemann tensor in (3.14) will ultimately yield the trace of X_{ABCD}.

It is well known that with the aid of the Kronecker delta

$$\begin{aligned} R_{a[bcd]} &= \delta_{[b}{}^{e}\delta_{c}{}^{f}\delta_{d]}{}^{g}R_{aefg} \\ &= -\frac{1}{6}\epsilon_{hbcd}(\epsilon^{hefg}R_{aefg}) \\ &= -\frac{1}{3}\epsilon_{hbcd}R^{*}{}_{ae}{}^{he} \end{aligned} \tag{3.27}$$

where the *right-dual* of the Riemann tensor is defined as

$$R^{*}{}_{abcd} = \frac{1}{2}\epsilon_{cd}{}^{ef}R_{abef}. \tag{3.28}$$

Hence, comparing (3.27) with (3.15) implies that

$$R^{*}{}_{ab}{}^{cb} = 0. \tag{3.29}$$

The spinor equivalent of (3.28) can be obtained by employing (2.129). Thus,

$$\begin{aligned} R^{*}{}_{abcd} &= \frac{i}{2}(\epsilon_{C}{}^{E}\epsilon_{D}{}^{F}\epsilon_{C'}{}^{F'}\epsilon_{D'}{}^{E'} - \epsilon_{C}{}^{F}\epsilon_{D}{}^{E}\epsilon_{C'}{}^{E'}\epsilon_{D'}{}^{F'})R_{AA'BB'EE'FF'} \\ &= iR_{AA'BB'CD'DC'}. \end{aligned} \tag{3.30}$$

Interchanging C' and D' in (3.20) and substituting this into (3.30) yields

$$\begin{aligned} R^{*}{}_{abcd} = &- iX_{ABCD}\epsilon_{A'B'}\epsilon_{C'D'} + i\Phi_{ABC'D'}\epsilon_{A'B'}\epsilon_{CD} \\ &- i\overline{\Phi}_{A'B'CD}\epsilon_{AB}\epsilon_{C'D'} + i\overline{X}_{A'B'C'D'}\epsilon_{AB}\epsilon_{CD}. \end{aligned} \tag{3.31}$$

Raising the index d and contracting with b implies, with the aid of (3.29), that

$$X_{ABC}{}^{B}\epsilon_{A'C'} = \overline{X}_{A'B'C'}{}^{B'}\epsilon_{AC}. \tag{3.32}$$

Notice that on contraction of d, b the second and third term on the right of (3.31) cancel with each other when the symmetries (3.24) and (3.26) are observed.

Transvecting through (3.32) by $\epsilon^{A'C'}$ gives

$$X_{ABC}{}^{B} = \frac{1}{2}\overline{X}_{A'B'}{}^{A'B'}\epsilon_{AC}. \tag{3.33}$$

Transvecting through this by ϵ^{AC} yields

$$\Lambda = \overline{\Lambda} \tag{3.34}$$

where the real scalar quantity Λ is defined by

$$\Lambda = \frac{1}{6}X_{AB}{}^{AB}. \tag{3.35}$$

On substituting (3.35) into (3.33) we obtain

$$X_{ABC}{}^{B} = 3\Lambda\epsilon_{AC}, \tag{3.36}$$

thus establishing $X_{ABC}{}^{B}$ to be skew-symmetric in the first and third indices. This can also be seen via the symmetries given by (3.23) and (3.25). Thence,

$$X_{ABC}{}^{B} = -X_{CBA}{}^{B}. \tag{3.37}$$

3.2.2 *The Ricci spinor*

The spinor equivalent of the Ricci tensor is obtained by transvecting through (3.20) by $\epsilon^{BD}\epsilon^{B'D'}$:

$$\begin{aligned} R_{ac} = R_{AA'CC'} = X_{ABC}{}^{B}\epsilon_{A'C'} - \Phi_{ACC'A'} \\ -\overline{\Phi}_{A'C'CA} + \overline{X}_{A'B'C'}{}^{B'}\epsilon_{AC}, \end{aligned} \tag{3.38}$$

where, evidently, we have chosen the Ricci tensor as

$$R_{ac} = R_{ab}{}^{b}{}_{c}. \tag{3.39}$$

Utilising the relations (3.26) and (3.36) we can rewrite (3.38) as

$$R_{ac} = R_{AA'CC'} = 6\Lambda\epsilon_{AC}\epsilon_{A'C'} - 2\Phi_{ACA'C'}. \tag{3.40}$$

In this context one frequently refers to $\Phi_{ABC'D'}$ as the *Ricci spinor.*

On transvection through by $\epsilon^{AC}\epsilon^{A'C'}$ (3.40) yields

$$R = R_{a}{}^{a} = 24\Lambda - \Phi_{A}{}^{A}{}_{A'}{}^{A'}. \tag{3.41}$$

However, the symmetry (3.24) ensures that $\Phi_{ABC'D'}$ has vanishing trace. Hence,

$$R = 24\Lambda. \tag{3.42}$$

It is now a simple matter to obtain the spinor equivalent of the trace-free Ricci tensor $R_{ab} - \frac{1}{4}Rg_{ab}$. Thus

$$R_{ab} - \frac{1}{4}Rg_{ab} = -2\Phi_{ABA'B'}. \tag{3.43}$$

The spinor equivalent of the Einstein tensor, $G_{ab} = R_{ab} - \frac{1}{2}Rg_{ab}$, is similarly obtained:

$$G_{ABA'B'} = -6\Lambda\epsilon_{AB}\epsilon_{A'B'} - 2\Phi_{ABA'B'}. \tag{3.44}$$

The full Einstein field equations are given by

$$G_{ab} = -8\pi\gamma T_{ab} - \lambda g_{ab} \tag{3.45}$$

where the velocity of light is taken as unity. The quantities T_{ab}, γ and λ are respectively the energy-momentum tensor, Newton's gravitational constant, and the cosmological constant. The spinor form of (3.45) is given by

$$-6\Lambda\epsilon_{AB}\epsilon_{A'B'} - 2\Phi_{ABA'B'} = -8\pi\gamma T_{ABA'B'} - \lambda\epsilon_{AB}\epsilon_{A'B'}. \tag{3.46}$$

On transvecting through by $\epsilon^{AB}\epsilon^{A'B'}$ we obtain

$$\Lambda = \frac{1}{3}\pi\gamma T_A{}^A{}_{A'}{}^{A'} + \frac{1}{6}\lambda. \tag{3.47}$$

Substituting this into (3.46) yields

$$\Phi_{ABA'B'} = 4\pi\gamma(T_{ABA'B'} - \frac{1}{4}T_X{}^X{}_{X'}{}^{X'}\epsilon_{AB}\epsilon_{A'B'}). \tag{3.48}$$

If one invokes the empty space field equations with non-zero cosmological constant, equations (3.47) and (3.48) are given respectively by

$$\Lambda = \frac{1}{6}\lambda \tag{3.49}$$

and

$$\Phi_{ABA'B'} = 0. \tag{3.50}$$

The empty space field equations with zero cosmological constant lead to

$$\Phi_{ABA'B'} = 0 = \Lambda. \tag{3.51}$$

Notice that $\Lambda = 0$ implies from (3.36) that

$$X_{A[B|C|D]} = 0, \tag{3.52}$$

and, hence, from (3.23)

$$X_{ABCD} = X_{(ABCD)}. \tag{3.53}$$

3.2.3 *The Weyl spinor*

In Sec. 2.6 our discussion concerned a particular symmetry property inherent in spinor structure. Namely, that *all* spinors with skew-symmetric components can be replaced with outer products of Levi–Civita spinors and lower valence spinors.

Let us assume Λ non-vanishing, and decompose the curvature spinor X_{ABCD} into parts which are totally symmetric and totally skew-symmetric plus additional terms that balance the equation:

$$\begin{aligned} X_{ABCD} = {} & X_{(ABCD)} + X_{[ABCD]} \\ & + \frac{11}{12} X_{ABCD} - \frac{1}{12}(X_{ACDB} + X_{ADBC} + X_{BCAD} + X_{BADC} \\ & + X_{BDCA} + X_{CBDA} + X_{CDAB} + X_{CABD} + X_{DACB} \\ & + X_{DCBA} + X_{DBAC}). \end{aligned} \tag{3.54}$$

Because of the symmetries (3.23) and (3.25) we have $X_{[ABCD]} = 0$. The same two symmetries also imply that

$$X_{(ABCD)} = X_{A(BCD)}. \tag{3.55}$$

Then (3.54) can be reduced to

$$X_{ABCD} = X_{A(BCD)} + \frac{1}{3}(X_{ABCD} - X_{ACBD}) + \frac{1}{3}(X_{ABCD} - X_{ADCB}). \tag{3.56}$$

By employing (2.57) we obtain

$$X_{ABCD} = X_{A(BCD)} + \frac{1}{3}\epsilon_{BC} X_{AE}{}^{E}{}_{D} + \frac{1}{3}\epsilon_{BD} X_{AEC}{}^{E}, \tag{3.57}$$

which, with the aid of (3.36), can be written as

$$X_{ABCD} = \Psi_{ABCD} + \Lambda(\epsilon_{BC}\epsilon_{AD} + \epsilon_{BD}\epsilon_{AC}) \tag{3.58}$$

where we define

$$\Psi_{ABCD} = X_{A(BCD)} = X_{ABCD} \tag{3.59}$$

to be the *Weyl (conformal) spinor* or *gravitational spinor*. Indeed, it was this spinor that we classified in Sec. 2.9.2 according to Petrov's scheme.

Clearly, because the Weyl spinor is purely symmetric, it possesses five mutually independent complex components.

It is now a simple matter to express R_{abcd} in terms of the two gravitational spinors Ψ_{ABCD}, $\Phi_{ABC'D'}$, and the real scalar Λ. Hence, from (3.20) we have

$$\begin{aligned} R_{abcd} &= \Psi_{ABCD}\epsilon_{A'B'}\epsilon_{C'D'} + \overline{\Psi}_{A'B'C'D'}\epsilon_{AB}\epsilon_{CD} \\ &+\Phi_{ABC'D'}\epsilon_{A'B'}\epsilon_{CD} + \overline{\Phi}_{A'B'CD}\epsilon_{AB}\epsilon_{C'D'} \\ &+2\Lambda(\epsilon_{AC}\epsilon_{BD}\epsilon_{A'C'}\epsilon_{B'D'} - \epsilon_{AD}\epsilon_{BC}\epsilon_{A'D'}\epsilon_{B'C'}). \end{aligned} \tag{3.60}$$

The curvature spinor $\Phi_{ABC'D'}$ possesses three complex and three real components and Λ has one real component. Together with the five complex components of the Weyl spinor, (3.60) exhibits the total sum of 20 real components of the Riemann tensor.

Aside

By defining

$$C_{abcd} = \Psi_{ABCD}\epsilon_{A'B'}\epsilon_{C'D'} + \overline{\Psi}_{A'B'C'D'}\epsilon_{AB}\epsilon_{CD} \tag{3.61}$$

it is now possible to show that the last column in Table 2.9.2 is equivalent to the last column in Table 2.9.3. (See Exercise 2). Also, from the dual of the Weyl tensor

$$C^*{}_{abcd} = \frac{1}{2}\epsilon_{cd}{}^{ef}C_{abef}, \tag{3.62}$$

we can define the *anti-self-dual* as (see (2.136))

$$\begin{aligned} -C_{abcd} &= \frac{1}{2}(C_{abcd} + iC^*{}_{abcd}) & (3.63) \\ &= \Psi_{ABCD}\epsilon_{A'B'}\epsilon_{C'D'}. & (3.64) \end{aligned}$$

It is now simple to see the equivalence between the Weyl tensor and spinor, and that replacing C_{abcd} with (3.63) in Table 2.9.2 will leave the conditions invariant. (Note that in (3.62) the left and right duals of C_{abcd} are equivalent.)

3.3 Spinor equivalent of the Ricci identities

It is common knowledge that when a commutator $2\nabla_{[a}\nabla_{b]}$ operates on an arbitrary vector or tensor field the outcome is a relation involving the Riemann tensor (see Appendix). A similar scenario arises when *spinor commutation operators* act on arbitrary univalent spinor fields.

Let us first obtain the spinor equivalent of $2\nabla_{[a}\nabla_{b]}$. We have

$$2\nabla_{[a}\nabla_{b]} = \nabla_{AA'}\nabla_{BB'} - \nabla_{BB'}\nabla_{AA'}. \tag{3.65}$$

By employing (2.60) and applying it to the index pair A, B we find that

$$\begin{aligned}2\nabla_{[a}\nabla_{b]} = {} & \nabla_{(A|A'|}\nabla_{B)B'} + \frac{1}{2}\epsilon_{AB}\nabla_{CA'}\nabla^{C}{}_{B'} \\ & -\nabla_{(B|B'|}\nabla_{A)A'} - \frac{1}{2}\epsilon_{BA}\nabla_{CB'}\nabla^{C}{}_{A'}.\end{aligned} \tag{3.66}$$

A second application of (2.60), this time on the index pair A', B', and necessary only on the first and third terms of (3.66), yields

$$\begin{aligned}2\nabla_{[a}\nabla_{b]} = {} & \nabla_{(A(A'}\nabla_{B)B')} + \frac{1}{2}\epsilon_{AB}\nabla_{CA'}\nabla^{C}{}_{B'} + \frac{1}{2}\epsilon_{A'B'}\nabla_{C'A}\nabla^{C'}{}_{B} \\ & -\nabla_{(B(B'}\nabla_{A)A')} - \frac{1}{2}\epsilon_{BA}\nabla_{CB'}\nabla^{C}{}_{A'} \\ & -\frac{1}{2}\epsilon_{B'A'}\nabla_{C'B}\nabla^{C'}{}_{A}.\end{aligned} \tag{3.67}$$

Note that in the first term and fourth term of (3.67) symmetrisation takes place between either pairs of primed or unprimed indices, but not mixed indices.

By observing all the symmetries of (3.67) we eventually obtain the relation

$$2\nabla_{[a}\nabla_{b]} = \epsilon_{A'B'}\square_{AB} + \epsilon_{AB}\square_{A'B'}, \tag{3.68}$$

where

$$\square_{AB} = \nabla_{C'(A}\nabla^{C'}{}_{B)}, \quad \square_{A'B'} = \nabla_{C(A'}\nabla^{C}{}_{B')}. \tag{3.69}$$

If one compares (3.68) with the decomposition (2.126) then

$$\overline{\square_{AB}} = \square_{A'B'}. \tag{3.70}$$

Notice the absence of a bar above the operator on the right-hand side of (3.70).

For an arbitrary vector field the *Ricci identity* can be written as

$$2\nabla_{[a}\nabla_{b]}V^d = R_{abc}{}^dV^c, \qquad (3.71)$$

where the Riemann tensor is defined in terms of the *Christoffel symbols*, $\Gamma^{\mathbf{a}}_{\mathbf{bc}}$, and their derivatives by

$$R_{\mathbf{abc}}{}^{\mathbf{d}} = \partial_{\mathbf{a}}\Gamma^{\mathbf{d}}_{\mathbf{cb}} - \partial_{\mathbf{b}}\Gamma^{\mathbf{d}}_{\mathbf{ca}} + \Gamma^{\mathbf{e}}_{\mathbf{cb}}\Gamma^{\mathbf{d}}_{\mathbf{ea}} - \Gamma^{\mathbf{e}}_{\mathbf{ca}}\Gamma^{\mathbf{d}}_{\mathbf{eb}}. \qquad (3.72)$$

Let us now introduce a four-dimensional basis $\delta^a_{\mathbf{a}}$ (arbitrarily chosen) such that

$$V^a = V^{\mathbf{a}}\delta^a_{\mathbf{a}} \qquad (3.73)$$

for any vector V^a. Recall (see Sec. 2.3) that boldface indices are numerical and label the vectors; whereas non-boldface indices label the basis elements. Note also that $V^{\mathbf{a}}$ is *not* a vector, but scalar components.

Substituting (3.73) into (3.71) yields

$$2V^{\mathbf{d}}\nabla_{[a}\nabla_{b]}\delta^d_{\mathbf{d}} = R_{abc}{}^dV^c, \qquad (3.74)$$

or

$$2(V^c\delta^{\mathbf{d}}_c)\nabla_{[a}\nabla_{b]}\delta^d_{\mathbf{d}} = R_{abc}{}^dV^c. \qquad (3.75)$$

Notice that the equation $\nabla_{[a}\nabla_{b]}V^{\mathbf{d}}$ is necessarily zero due to $V^{\mathbf{d}}$ being a scalar field. From (2.73) we obtain

$$2\delta^{\mathbf{d}}_c\nabla_{[a}\nabla_{b]}\delta^d_{\mathbf{d}} = R_{abc}{}^d. \qquad (3.76)$$

The spinor equivalent of (3.76) can be obtained by employing (2.84), (3.15) and (3.68). Thence,

$$\epsilon_C{}^{\mathbf{D}}\epsilon_{C'}{}^{\mathbf{D}'}[\epsilon_{A'B'}\Box_{AB} + \epsilon_{AB}\Box_{A'B'}]\epsilon_{\mathbf{D}}{}^D\epsilon_{\mathbf{D}'}{}^{D'} = R_{AA'BB'CC'}{}^{DD'}, \qquad (3.77)$$

or, in terms of the curvature spinors given by (3.19)

$$\begin{aligned}\epsilon_C{}^{\mathbf{D}}\epsilon_{C'}{}^{D'}[\epsilon_{A'B'}\Box_{AB}\epsilon_{\mathbf{D}}{}^D &+ \epsilon_{AB}\Box_{A'B'}\epsilon_{\mathbf{D}}{}^D] \\ &+ \epsilon_{C'}{}^{\mathbf{D}'}\epsilon_C{}^D[\epsilon_{A'B'}\Box_{AB}\epsilon_{\mathbf{D}'}{}^{D'} + \epsilon_{AB}\Box_{A'B'}\epsilon_{\mathbf{D}'}{}^{D'}] \\ = X_{ABC}{}^D\epsilon_{A'B'}\epsilon_{C'}{}^{D'} &+ \Phi_{ABC'}{}^{D'}\epsilon_{A'B'}\epsilon_C{}^D \\ &+ \overline{\Phi}_{A'B'C}{}^D\epsilon_{AB}\epsilon_{C'}{}^{D'} + \overline{X}_{A'B'C'}{}^{D'}\epsilon_{AB}\epsilon_C{}^D.\end{aligned} \qquad (3.78)$$

It follows, lowering the indices D and D', that

$$\epsilon_C{}^{\mathbf{D}'}\Box_{AB}\epsilon_{\mathbf{D}D} = X_{ABCD} \qquad (3.79)$$

and

$$\epsilon_C{}^{\mathbf{D}}\Box_{A'B'}\epsilon_{\mathbf{D}D} = \Phi_{A'B'CD}. \tag{3.80}$$

Similar expressions involving the Weyl spinor Ψ_{ABCD} and the scalar Λ can be obtained by substituting (3.58) into (3.79):

$$\epsilon_C{}^{\mathbf{D}}\Box_{AB}\epsilon_{\mathbf{D}D} = \Psi_{ABCD} + \Lambda(\epsilon_{BC}\epsilon_{AD} + \epsilon_{BD}\epsilon_{AC}). \tag{3.81}$$

Symmetrising on the indices A, B, C implies that

$$\epsilon_C{}^{\mathbf{D}}\Box_{(AB}\epsilon_{|\mathbf{D}|D)} = \Psi_{ABCD}. \tag{3.82}$$

Transvecting through (3.81) by $\epsilon^{BC}\epsilon^{AD}$ yields

$$\frac{1}{6}\epsilon_C{}^{\mathbf{D}}\Box^{CD}\epsilon_{\mathbf{D}D} = \Lambda. \tag{3.83}$$

Effectively, replacing $\epsilon_{\mathbf{D}D}$ by some arbitrary univalent spinor ζ_D, say, means that we can write expressions (3.79), (3.80), (3.81), (3.82) and (3.83) respectively as

$$\Box_{AB}\zeta_D = X_{ABDC}\zeta^C, \tag{3.84}$$

$$\Box_{AB}\zeta_D = \Phi_{A'B'DC}\zeta^C, \tag{3.85}$$

$$\Box_{AB}\zeta_D = \Psi_{ABDC}\zeta^C - \Lambda(\epsilon_{BD}\zeta_A + \epsilon_{AD}\zeta_B), \tag{3.86}$$

$$\Box_{(AB}\zeta_{D)} = \Psi_{ABDC}\zeta^C, \tag{3.87}$$

$$\Box^{AB}\zeta_B = 3\Lambda\zeta^A. \tag{3.88}$$

The equations (3.84)–(3.88) are called the *spinor equivalents of the Ricci identities.*

3.4 Spinor equivalent of the Bianchi identities

Consider the Bianchi identities in tensor form; namely,

$$\nabla_{[a}R_{bc]de} = 0. \tag{3.89}$$

It will be more convenient to work with the Bianchi identities in a form that exhibits the dual of the Riemann tensor. Thus we can write

$$\begin{aligned}\nabla_{[a}R_{bc]de} &= \delta_{[a}{}^{p}\delta_{b}{}^{q}\delta_{c]}{}^{r}\nabla_{p}R_{qrde}\\ &= -\frac{1}{6}\epsilon_{abcs}\epsilon^{pqrs}\nabla_{p}R_{qrde}\\ &= -\frac{1}{3}\epsilon_{abcs}\nabla^{p*}R^{s}{}_{pde}. \end{aligned}\tag{3.90}$$

Hence from (3.87) we have

$$\nabla^{a*}R_{abcd} = 0. \tag{3.91}$$

Substituting (3.31) into (3.91), and remembering that the Levi–Civita symbol is covariantly constant, yields

$$\begin{aligned}-\epsilon_{C'D'}\nabla^{A}_{B'}X_{ABCD} - \epsilon_{CD}\nabla^{A}_{B'}\Phi_{ABC'D'}&\\ +\epsilon_{C'D'}\nabla^{A'}_{B}\overline{\Phi}_{A'B'CD} + \epsilon_{CD}\nabla^{A'}_{B}\overline{X}_{A'B'C'D'} &= 0.\end{aligned}\tag{3.92}$$

Symmetrising on C', D' gives

$$\nabla^{A'}_{B}\overline{X}_{A'B'C'D'} = \nabla^{A}_{B'}\Phi_{ABC'D'}. \tag{3.93}$$

Symmetrising on C, D gives

$$\nabla^{A}_{B'}X_{ABCD} = \nabla^{A'}_{B}\Phi_{A'B'CD}, \tag{3.94}$$

where we have used (3.26) in the last two equations. Equation (3.93), or (3.94), constitute the *spinor equivalent of the Bianchi identities.*

We can go a stage further by incorporating the Weyl spinor Ψ_{ABCD} and the scalar curvature Λ. Substituting (3.58) into (3.94) yields

$$\nabla^{A}_{B'}\Psi_{ABCD} + (\epsilon_{BC}\epsilon_{AD} + \epsilon_{BD}\epsilon_{AC})\nabla^{A}_{B'}\Lambda = \nabla^{A'}_{B}\Phi_{A'B'CD}, \tag{3.95}$$

which symmetrising on B, C and B, D gives

$$\nabla^{A}_{B'}\Psi_{ABCD} = \nabla^{A'}_{(B}\Phi_{CD)A'B'}. \tag{3.96}$$

If we now skew-symmetrise (3.95) on B, C and then transvect through by ϵ^{CB} we obtain the contracted Bianchi identities:

$$\nabla^{CA'}\Phi_{CDA'B'} = -3\nabla_{DB'}\Lambda. \tag{3.97}$$

Notice that if one were to covariantly differentiate equation (3.44) and compare this with (3.97) then

$$\nabla^{AA'} G_{AA'BB'} = 0, \tag{3.98}$$

that is, the Einstein tensor is divergence-free.

3.5 The Newman–Penrose spin coefficient formalism

Consider a set of normalised, linearly independent tetrad vector fields chosen at each point of space-time. (The point may perhaps be arbitrarily chosen if there is little structure defined by the geometry at that point.) We can now translate any tensor equation into a system of scalar equations by taking components with respect to the tetrad vectors in question, and performing appropriate calculations with them.

Effectively, the same situation occurs when dealing with the *Newman–Penrose spin coefficient formalism* (Newman and Penrose, 1962) (usually abbreviated to *NP formalism*). In contrast, however, instead of employing an orthonormal basis, the complex null basis of (2.95) is utilised. Furthermore, instead of the 24 independent real scalar quantities associated with the orthodox normalised tetrad formalism, one need only to consider 12 independent complex scalar quantities during explicit calculations. Thus we have acquired a rather economical formalism with a significant reduction in components.

3.5.1 *Spin coefficients*

Consider the expression

$$\nabla_{AA'} \zeta^B. \tag{3.99}$$

Then for $\zeta^B \in S^B$ we have components

$$\zeta^{\mathbf{B}} = \zeta^B \epsilon_B{}^{\mathbf{B}}. \tag{3.100}$$

We wish to project (3.99) on to a dyad basis

$$\epsilon_{\mathbf{A}}{}^A = (o^A, \iota^A) \tag{3.101}$$

where $\mathbf{A}$ represents the dyad indices $0, 1$ and A is the abstract index such that $\epsilon_{\mathbf{A}}{}^A \in S^A$, the dual basis being $\epsilon_A{}^{\mathbf{A}}$. To do this we write the dyad form of $\nabla_{AA'}$. Thus,

$$\epsilon_{\mathbf{A}}{}^{A}\epsilon_{\mathbf{A}'}{}^{A'}\nabla_{AA'} = \nabla_{\mathbf{AA}'} \tag{3.102}$$

where, of course, $\epsilon_{\mathbf{A}'}{}^{A'}$ is the complex conjugate of $\epsilon_{\mathbf{A}}{}^{A}$. Now replace $\nabla_{AA'}$ on the left-hand side of (3.102) with (3.99), and replace $\nabla_{\mathbf{AA}'}$ on the right-hand side with $\nabla_{\mathbf{AA}'}(\zeta^{\mathbf{C}}\epsilon_{\mathbf{C}}{}^{B})$. If we also transvect through by $\epsilon_{B}{}^{\mathbf{B}}$, (3.102) becomes

$$\begin{aligned}
\epsilon_{\mathbf{A}}{}^{A}\epsilon_{\mathbf{A}'}{}^{A'}\epsilon_{B}{}^{\mathbf{B}}\nabla_{AA'}\zeta^{B} &= \epsilon_{B}{}^{\mathbf{B}}\nabla_{\mathbf{AA}'}(\zeta^{\mathbf{C}}\epsilon_{\mathbf{C}}{}^{B}) \\
&= \epsilon_{\mathbf{C}}{}^{B}\epsilon_{B}{}^{\mathbf{B}}\nabla_{\mathbf{AA}'}\zeta^{\mathbf{C}} + \zeta^{\mathbf{C}}\epsilon_{B}{}^{\mathbf{B}}\nabla_{\mathbf{AA}'}\epsilon_{\mathbf{C}}{}^{B} \\
&= \nabla_{\mathbf{AA}'}\zeta^{\mathbf{B}} + \zeta^{\mathbf{C}}\gamma_{\mathbf{AA}'\mathbf{C}}{}^{\mathbf{B}},
\end{aligned} \tag{3.103}$$

where we define

$$\gamma_{\mathbf{AA}'\mathbf{C}}{}^{\mathbf{B}} = \epsilon_{B}{}^{\mathbf{B}}\nabla_{\mathbf{AA}'}\epsilon_{\mathbf{C}}{}^{B}. \tag{3.104}$$

Because the Kronecker spinor is covariantly constant, we can obtain by direct calculation

$$\nabla_{\mathbf{AA}'}\epsilon_{\mathbf{C}}{}^{\mathbf{B}} = \nabla_{\mathbf{AA}'}(\epsilon_{A}{}^{\mathbf{B}}\epsilon_{\mathbf{C}}{}^{A}) = \epsilon_{A}{}^{\mathbf{B}}\nabla_{\mathbf{AA}'}\epsilon_{\mathbf{C}}{}^{A} + \epsilon_{\mathbf{C}}{}^{A}\nabla_{\mathbf{AA}'}\epsilon_{A}{}^{\mathbf{B}} = 0,$$

implying from (3.104) that:

$$\gamma_{\mathbf{AA}'\mathbf{C}}{}^{\mathbf{B}} = -\epsilon_{\mathbf{C}}{}^{B}\nabla_{\mathbf{AA}'}\epsilon_{B}{}^{\mathbf{B}}. \tag{3.105}$$

Also, the complex conjugate of $\gamma_{\mathbf{AA}'\mathbf{C}}{}^{\mathbf{B}}$ is simply

$$\overline{\gamma}_{\mathbf{AA}'\mathbf{C}'}{}^{\mathbf{B}'} = \epsilon_{B'}{}^{\mathbf{B}'}\nabla_{\mathbf{AA}'}\epsilon_{\mathbf{C}'}{}^{B} = -\epsilon_{\mathbf{C}'}{}^{B'}\nabla_{\mathbf{AA}'}\epsilon_{B'}{}^{\mathbf{B}'}. \tag{3.106}$$

The quantities $\gamma_{\mathbf{AA}'\mathbf{C}}{}^{\mathbf{B}}$ are called *spinor Ricci rotation coefficients.* They define 12 independent complex functions with respect to a normalised spin-basis at each point of space-time. If the spin-basis is not normalised a total of 16 scalar quantities are obtainable. The 12 (or 16) explicit forms of the Ricci rotation coefficients are known as *spin coefficients* and are frequently utilised in general relativistic calculations; particularly those involving exact solutions of Einstein's field equations. Let us repeat the procedure which led to (3.103), only this time $\nabla_{\mathbf{AA}'}$ is allowed to operate on the covariant univalent spinor η_B. Since the calculation is straightforward, we quote only the result:

$$\epsilon_{\mathbf{A}}{}^{A}\epsilon_{\mathbf{A}'}{}^{A'}\epsilon_{\mathbf{B}}{}^{B}\nabla_{AA'}\eta_{B} = \nabla_{\mathbf{AA}'}\eta_{\mathbf{B}} - \eta_{\mathbf{C}}\gamma_{\mathbf{AA}'\mathbf{B}}{}^{\mathbf{C}}. \tag{3.107}$$

The complex conjugates of (3.103) and (3.107) are, respectively,

$$\epsilon_{\mathbf{A}}{}^{A}\epsilon_{\mathbf{A}'}{}^{A'}\epsilon_{B'}{}^{\mathbf{B}}\nabla_{AA'}\mu^{B'} = \nabla_{\mathbf{AA}'}\mu^{\mathbf{B}'} + \mu^{\mathbf{C}'}\overline{\gamma}_{\mathbf{AA}'C'}{}^{\mathbf{B}} \tag{3.108}$$

and

$$\epsilon_{\mathbf{A}}{}^{A}\epsilon_{\mathbf{A}'}{}^{A'}\epsilon_{\mathbf{B}'}{}^{B}\nabla_{AA'}\nu_{B'} = \nabla_{\mathbf{AA'}}\nu_{\mathbf{B'}} - \nu_{\mathbf{C'}}\overline{\gamma}_{\mathbf{AA'B'}}{}^{\mathbf{C'}} \tag{3.109}$$

where $\mu^{B'} \equiv \overline{\zeta^B}$ and $\nu_{B'} \equiv \overline{\eta_B}$. Note that the relations (3.108) and (3.109) are equivalent to obtaining the components of $\nabla_{AA'}\mu^{B'}$ and $\nabla_{AA'}\nu_{B'}$ via (3.103) and (3.107). And for any $(p,q;r,s)$ spinor (i.e. a general multivalent spinor) $\chi^{B\cdots C'\cdots}_{D\cdots E'\cdots}$ the components of $\nabla_{AA'}\chi^{B\cdots C'\cdots}_{D\cdots E'\cdots}$ are:

$$\begin{aligned}
\epsilon_{\mathbf{A}}{}^{A}\epsilon_{\mathbf{A}'}{}^{A'}\epsilon_{B}{}^{\mathbf{B}}\cdots\epsilon_{C'}{}^{\mathbf{C}'}\cdots\epsilon_{\mathbf{D}}{}^{D}\cdots\epsilon_{\mathbf{E}'}{}^{E'}\cdots\nabla_{AA'}\chi^{B\cdots C'\cdots}_{D\cdots E'\cdots}& \\
= \epsilon_{B}{}^{\mathbf{B}}\cdots\epsilon_{C'}{}^{\mathbf{C}'}\cdots\epsilon_{\mathbf{D}}{}^{D}\cdots\epsilon_{\mathbf{E}'}{}^{E'}\cdots& \\
\nabla_{\mathbf{AA'}}(\chi^{\mathbf{W}\cdots\mathbf{X}'\cdots}_{\mathbf{Y}\cdots\mathbf{Z}'\cdots}\epsilon_{\mathbf{W}}{}^{B}\cdots\epsilon_{\mathbf{X}'}{}^{C'}\cdots\epsilon_{D}{}^{\mathbf{Y}}\cdots\epsilon_{E'}{}^{\mathbf{Z}'}\cdots)& \\
= \nabla_{\mathbf{AA'}}\chi^{\mathbf{B}\cdots\mathbf{C}'\cdots}_{\mathbf{D}\cdots\mathbf{E}'\cdots} + \chi^{\mathbf{W}\cdots\mathbf{C}'\cdots}_{\mathbf{D}\cdots\mathbf{E}'\cdots}\gamma_{\mathbf{AA'W}}{}^{\mathbf{B}} + \cdots& \\
+\chi^{\mathbf{B}\cdots\mathbf{X}'\cdots}_{\mathbf{D}\cdots\mathbf{E}'\cdots}\overline{\gamma}_{\mathbf{AA'X'}}{}^{\mathbf{C}'} + \cdots - \chi^{\mathbf{B}\cdots\mathbf{C}'\cdots}_{\mathbf{Y}\cdots\mathbf{E}'\cdots}\gamma_{\mathbf{AA'D}}{}^{\mathbf{Y}} - \cdots& \\
-\chi^{\mathbf{B}\cdots\mathbf{C}'\cdots}_{\mathbf{D}\cdots\mathbf{Z}'\cdots}\overline{\gamma}_{\mathbf{AA'E'}}{}^{\mathbf{Z}'} - \cdots.&
\end{aligned} \tag{3.110}$$

That the Ricci rotation coefficients are symmetric in the last two indices can be shown by allowing $\nabla_{\mathbf{AA'}}$ to operate on the Kronecker spinor, which, of course, is covariantly constant. Thence,

$$\begin{aligned}
\epsilon_{\mathbf{A}}{}^{A}\epsilon_{\mathbf{A}'}{}^{A'}\epsilon_{C}{}^{\mathbf{C}}\epsilon_{\mathbf{B}}{}^{B}\nabla_{AA'}\epsilon_{B}{}^{C} &= \epsilon_{C}{}^{\mathbf{C}}\epsilon_{\mathbf{B}}{}^{B}\nabla_{\mathbf{AA'}}(\epsilon_{\mathbf{D}}{}^{C}\epsilon_{B}{}^{\mathbf{D}}) \\
&= \epsilon^{B}{}_{\mathbf{B}}\nabla_{\mathbf{AA'}}\epsilon^{\mathbf{C}}{}_{B} + \epsilon_{c}{}^{\mathbf{C}}\nabla_{\mathbf{AA'}}\epsilon_{\mathbf{B}}{}^{C} \\
&= -\gamma_{\mathbf{AA'}}{}^{\mathbf{C}}{}_{\mathbf{B}} + \gamma_{\mathbf{AA'B}}{}^{\mathbf{C}} \\
&= 0,
\end{aligned}$$

implying that

$$\gamma_{\mathbf{AA'BC}} = \gamma_{\mathbf{AA'CB}}. \tag{3.111}$$

Had our spin basis not been normalised then the components of ϵ_{AB}, with respect to the basis $\epsilon_{\mathbf{A}}{}^{A} \in S^A$, shown in (2.32), would be given by

$$\epsilon_{\mathbf{AB}} = \epsilon_{AB}\epsilon_{\mathbf{A}}{}^{A}\epsilon_{\mathbf{B}}{}^{B} = \begin{pmatrix} 0 & \Omega \\ -\Omega & 0 \end{pmatrix} \tag{3.112}$$

where

$$\Omega = \epsilon_{AB}o^{A}\iota^{B} = o_{A}\iota^{A}. \tag{3.113}$$

This would mean that the symmetry exhibited in (3.111) would not hold.

For identification purposes it is conventional to assign Greek letters to each independent Ricci rotation coefficient. These spin coefficients are directly related to the Ricci rotation coefficients by

$$
\begin{aligned}
\epsilon &= \gamma_{00'0}{}^{0} = -\gamma_{00'1}{}^{1} = \gamma_{00'10} \\
\alpha &= \gamma_{10'0}{}^{0} = -\gamma_{10'1}{}^{1} = \gamma_{10'10} \\
\beta &= \gamma_{01'0}{}^{0} = -\gamma_{01'1}{}^{1} = \gamma_{01'10} \\
\gamma &= \gamma_{11'0}{}^{0} = -\gamma_{11'1}{}^{1} = \gamma_{11'10} \\
\pi &= \gamma_{00'1}{}^{0} = \gamma_{00'11} \\
\lambda &= \gamma_{10'1}{}^{0} = \gamma_{10'11} \\
\mu &= \gamma_{01'1}{}^{0} = \gamma_{01'11} \\
\nu &= \gamma_{11'1}{}^{0} = \gamma_{11'11} \\
\kappa &= -\gamma_{00'0}{}^{1} = \gamma_{00'00} \\
\rho &= -\gamma_{10'0}{}^{1} = \gamma_{10'00} \\
\sigma &= -\gamma_{01'0}{}^{1} = \gamma_{01'00} \\
\tau &= -\gamma_{11'0}{}^{1} = \gamma_{11'00}.
\end{aligned}
\tag{3.114}
$$

We now have at our disposal another way of describing the spin coefficients in (3.114). By employing (3.104) each spin coefficient can be described in terms of a contraction between a basis spinor (i.e. $o^A, o_A, \iota^A, \iota_A$) and the directional derivative of a basis spinor (i.e. $\nabla_{00'}o^A$ etc.). Alternatively, a contraction between a null tetrad vector (l^a, l_a, n^a, n_a etc.) and the directional derivative of a null tetrad vector (i.e. $\nabla_{00'}l^a$ etc.) will achieve an equivalent result. Thus with the aid of (2.32), (2.48), (2.94), (2.95) and (2.96) we have the following scheme:

$$
\begin{aligned}
\epsilon &= -\frac{1}{2}(l^a D n_a - m^a D\overline{m}_a) = \frac{1}{2}(n^a D l_a - \overline{m}^a D m_a) = \iota^A D o_A = o^A D\iota_A \\
\alpha &= -\frac{1}{2}(l^a \overline{\delta} n_a - m^a \overline{\delta}\overline{m}_a) = \frac{1}{2}(n^a \overline{\delta} l_a - \overline{m}^a \overline{\delta} m_a) = \iota^A \overline{\delta} o_A = o^A \overline{\delta}\iota_A \\
\beta &= -\frac{1}{2}(l^a \delta n_a - m^a \delta\overline{m}_a) = \frac{1}{2}(n^a \delta l_a - \overline{m}^a \delta m_a) = \iota^A \delta o_A = o^A \delta\iota_A \\
\gamma &= -\frac{1}{2}(l^a \Delta n_a - m^a \Delta\overline{m}_a) = \frac{1}{2}(n^a \Delta l_a - \overline{m}^a \Delta m_a) = \iota^A \Delta o_A = o^A \Delta\iota_A \\
\pi &= -\overline{m}^a D n_a = \iota^A D\iota_A \\
\lambda &= -\overline{m}^a \overline{\delta} n_a = \iota^A \overline{\delta}\iota_A \\
\mu &= -\overline{m}^a \delta n_a = \iota^A \delta\iota_A \\
\nu &= -\overline{m}^a \Delta n_a = \iota^A \Delta\iota_A \\
\kappa &= m^a D l_a = o^A D o_A \\
\rho &= m^a \overline{\delta} l_a = o^A \overline{\delta} o_A
\end{aligned}
$$

$$\sigma = m^a \delta l_a = o^A \delta o_A$$
$$\tau = m^a \Delta l_a = o^A \Delta o_A, \tag{3.115}$$

where we have used the conventional symbols $D, \Delta, \delta, \overline{\delta}$, known as *intrinsic derivatives*, to represent the directional derivatives along the direction $\mathbf{l}, \mathbf{n}, \mathbf{m}, \overline{\mathbf{m}}$ respectively. The intrinsic derivatives are defined by

$$\begin{aligned} D &= \nabla_{00'} = o^A o^{A'} \nabla_{AA'} = l^a \nabla_a \\ \Delta &= \nabla_{11'} = \iota^A \iota^{A'} \nabla_{AA'} = n^a \nabla_a \\ \delta &= \nabla_{01'} = o^A \iota^{A'} \nabla_{AA'} = m^a \nabla_a \\ \overline{\delta} &= \nabla_{10'} = \iota^A o^{A'} \nabla_{AA'} = \overline{m}^a \nabla_a. \end{aligned} \tag{3.116}$$

To illustrate the equivalence between the terms involving null tetrad vectors and spinor bases in (3.115), consider, for example, the spin coefficient π. Then

$$\begin{aligned} \pi &= -\overline{m}^a D n_a \\ &= -\iota^A o^{A'} D l_A l_{A'} \\ &= -\{\iota^A o^{A'} (\iota_A D \iota_{A'} + \iota_{A'} D \iota_A)\} \\ &= \iota^A D \iota_A. \end{aligned}$$

The equivalence between this and the corresponding Ricci rotation coefficient given in (3.114) is easily seen:

$$\pi = \gamma_{00'1}{}^0 = \epsilon_A{}^0 \nabla_{00'} \epsilon_1{}^A = \iota^A D \iota_A.$$

Note that proving the converse in the first procedure is more involved.

It is sometimes convenient in explicit calculations involving spin coefficients to exhibit (3.115) in a slightly different form. That is, each basis spinor o^A, ι^A can be operated on by each of the four intrinsic derivatives $D, \Delta, \delta, \overline{\delta}$, yielding combinations of basis spinors with spin coefficients. For example,

$$D o^A = \epsilon^{AB} D o_B, \tag{3.117}$$

which, with the aid of (2.45), gives

$$D o^A = o^A \epsilon - \iota^A \kappa.$$

The eight relations formed by applying the four intrinsic derivative operators to each basis spinor are given below:

$$D o^A = \epsilon o^A - \kappa \iota^A$$

$$\begin{aligned}
\Delta o^A &= \gamma o^A - \tau\iota^A \\
\delta o^A &= \beta o^A - \sigma\iota^A \\
\overline{\delta} o^A &= \alpha o^A - \rho\iota^A \\
D\iota^A &= \pi o^A - \epsilon\iota^A \\
\Delta\iota^A &= \nu o^A - \gamma\iota^A \\
\delta\iota^A &= \mu o^A - \beta\iota^A \\
\overline{\delta}\iota^A &= \lambda o^A - \alpha\iota^A .
\end{aligned} \tag{3.118}$$

The complex conjugates of these relations are evidently given by

$$\begin{aligned}
Do^{A'} &= \overline{\epsilon} o^{A'} - \overline{\kappa}\iota^{A'} \\
\Delta o^{A'} &= \overline{\gamma} o^{A'} - \overline{\tau}\iota^{A'} \\
\overline{\delta} o^{A'} &= \overline{\beta} o^{A'} - \overline{\sigma}\iota^{A'} \\
\delta o^{A'} &= \overline{\alpha} o^{A'} - \overline{\rho}\iota^{A'} \\
D\iota^{A'} &= \overline{\pi} o^{A'} - \overline{\epsilon}\iota^{A'} \\
\Delta\iota^{A'} &= \overline{\nu} o^{A'} - \overline{\gamma}\iota^{A'} \\
\overline{\delta}\iota^{A'} &= \overline{\mu} o^{A'} - \overline{\beta}\iota^{A'} \\
\delta\iota^{A'} &= \overline{\lambda} o^{A'} - \overline{\alpha}\iota^{A'} .
\end{aligned} \tag{3.119}$$

Using (3.118) will allow us to write similar expressions for the null tetrad vectors when operated upon by the intrinsic derivatives:

$$\begin{aligned}
Dl^a &= (\epsilon + \overline{\epsilon}) l^a - \overline{\kappa} m^a - \kappa \overline{m}^a \\
\Delta l^a &= (\gamma + \overline{\gamma}) l^a - \overline{\tau} m^a - \tau \overline{m}^a \\
\delta l^a &= (\beta + \overline{\alpha}) l^a - \overline{\rho} m^a - \sigma \overline{m}^a \\
\overline{\delta} l^a &= (\alpha + \overline{\beta}) l^a - \overline{\sigma} m^a - \rho \overline{m}^a \\
Dn^a &= -(\epsilon + \overline{\epsilon}) n^a + \pi m^a + \overline{\pi}\, \overline{m}^a \\
\Delta n^a &= -(\gamma + \overline{\gamma}) n^a + \nu \overline{m}^a + \overline{\nu}\, \overline{m}^a \\
\delta n^a &= -(\beta + \overline{\alpha}) n^a + \mu m^a + \overline{\lambda} \overline{m}^a \\
\overline{\delta} n^a &= -(\alpha + \overline{\beta}) n^a + \lambda m^a + \overline{\mu}\, \overline{m}^a \\
Dm^a &= (\epsilon - \overline{\epsilon}) m^a + \overline{\pi} l^a - \kappa n^a \\
\Delta m^a &= (\gamma - \overline{\gamma}) m^a + \overline{\nu} l^a - \tau n^a \\
\delta m^a &= (\beta - \overline{\alpha}) m^a + \overline{\lambda} l^a - \sigma n^a \\
\overline{\delta} m^a &= (\alpha - \overline{\beta}) m^a + \overline{\mu} l^a - \rho n^a \\
D\overline{m}^a &= (\overline{\epsilon} - \epsilon) \overline{m}^a + \pi l^a - \overline{\kappa} n^a
\end{aligned}$$

$$\begin{aligned}\Delta\overline{m}^a &= (\overline{\gamma}-\gamma)\overline{m}^a + \nu l^a - \overline{\tau} n^a\\ \delta\overline{m}^a &= (\overline{\alpha}-\beta)\overline{m}^a + \mu l^a - \overline{\rho} n^a\\ \overline{\delta}\overline{m}^a &= (\overline{\beta}-\alpha)\overline{m}^a + \lambda l^a - \overline{\sigma} n^a.\end{aligned} \tag{3.120}$$

3.5.2 *The Newman–Penrose field equations*

So far, we have not related the spin coefficients to any of the important spinor quantities in general relativity, such as the Weyl spinor, etc. However, this will be resolved presently by first obtaining the spin coefficient form of the *commutation relations*; that is, the commutator of the intrinsic derivatives given by (3.116).

The commutator (3.10) operating on a scalar quantity ϕ is given by

$$\{\nabla_a\nabla_b - \nabla_b\nabla_a\}\phi = 0. \tag{3.121}$$

Transvecting this with

$$A^a = \epsilon_{\mathbf{A}}{}^{A}\epsilon_{\mathbf{A}'}{}^{A'},\ B^b = \epsilon_{\mathbf{B}}{}^{B}\epsilon_{\mathbf{B}'}{}^{B'} \tag{3.122}$$

yields

$$B^b(A^a\nabla_a)\nabla_b\phi - A^a(B^b\nabla_b)\nabla_a\phi, \tag{3.123}$$

which can be written as

$$A^a\nabla_a(B^b\nabla_b\phi) - (A^a\nabla_a B^b)(\nabla_b\phi) - B^b\nabla_b(A^a\nabla_a\phi) + (B^b\nabla_b A^a)(\nabla_a\phi) = 0. \tag{3.124}$$

Replacing the appropriate quantities in (3.124) and (3.122) and rearranging gives

$$\begin{aligned}\nabla_{\mathbf{AA}'}\nabla_{\mathbf{BB}'}\phi - \nabla_{\mathbf{BB}'}\nabla_{\mathbf{AA}'}\phi = {}&\nabla_{\mathbf{AA}'}(\epsilon_{\mathbf{B}}{}^{B}\epsilon_{\mathbf{B}'}{}^{B})\nabla_{BB'}\phi\\ &-\nabla_{\mathbf{BB}'}(\epsilon_{\mathbf{A}'}{}^{A}\epsilon_{\mathbf{A}'}{}^{A'})\nabla_{AA'}\phi.\end{aligned} \tag{3.125}$$

We can write the two terms on the right-hand side of (3.125), using Liebniz's rule, as

$$\begin{aligned}&\epsilon_{\mathbf{B}}{}^{B}(\nabla_{\mathbf{AA}'}\epsilon_{\mathbf{B}'}{}^{B'})(\nabla_{BB'}\phi) + \epsilon_{\mathbf{B}}{}^{B'}(\nabla_{\mathbf{AA}'}\epsilon_{\mathbf{B}}{}^{B})(\nabla_{BB'}\phi)\\ &-\epsilon_{\mathbf{A}}{}^{A}(\nabla_{\mathbf{BB}'}\epsilon_{\mathbf{A}'}{}^{A'})(\nabla_{AA'}\phi) - \epsilon_{\mathbf{A}'}{}^{A'}(\nabla_{\mathbf{BB}'}\epsilon_{\mathbf{A}}{}^{A})(\nabla_{AA'}\phi).\end{aligned} \tag{3.126}$$

By applying (3.104) to the first term, say, of (3.126), we have

$$\epsilon_{\mathbf{B}}{}^{B}(\nabla_{\mathbf{AA}'}\epsilon_{\mathbf{B}'}{}^{B'})(\nabla_{BB'}\phi) = \overline{\gamma}_{\mathbf{AA}'B'}{}^{P'}\nabla_{\mathbf{BP}'}\phi.$$

Hence using (3.104) similarly with the other three terms in (3.126), we finally obtain from (3.125):

$$\nabla_{\mathbf{AA'}}\nabla_{\mathbf{BB'}}\phi - \nabla_{\mathbf{BB'}}\nabla_{\mathbf{AA'}}\phi = \gamma_{\mathbf{AA'B'}}{}^{\mathbf{P}}\nabla_{\mathbf{PB}}\phi - \gamma_{\mathbf{BB'A}}{}^{\mathbf{P}}\nabla_{\mathbf{PA'}}\phi$$
$$+\overline{\gamma}_{\mathbf{AA'B'}}{}^{\mathbf{P}}\nabla_{\mathbf{BP'}}\phi - \overline{\gamma}_{\mathbf{BB'A'}}{}^{\mathbf{P'}}\nabla_{\mathbf{AP'}}\phi. \tag{3.127}$$

It is now a straightforward exercise to interpret (3.127) explicitly in terms of the spin coefficients given in (3.114) and the intrinsic derivatives given in (3.116). We find that

$$\begin{aligned}
(\Delta D - D\Delta)\phi &= [(\gamma+\overline{\gamma})D + (\epsilon+\overline{\epsilon})\Delta - (\overline{\tau}+\pi)\delta - (\tau+\overline{\pi})\overline{\delta}]\phi\\
(\delta D - D\delta)\phi &= [(\overline{\alpha}+\beta-\overline{\pi})D + \kappa\Delta - (\overline{\rho}+\epsilon-\overline{\epsilon})\delta - \sigma\overline{\delta}]\phi\\
(\delta\Delta - \Delta\delta)\phi &= [-\overline{\nu}D - (\overline{\alpha}+\beta-\tau)\Delta + (\mu-\gamma+\overline{\gamma})\delta + \overline{\lambda\delta}]\phi\\
(\overline{\delta}\delta - \delta\overline{\delta})\phi &= [(\overline{\mu}-\mu)D + (\overline{\rho}-\rho)\Delta - (\overline{\beta}-\alpha)\delta - (\overline{\alpha}-\beta)\overline{\delta}]\phi.
\end{aligned} \tag{3.128}$$

The use of spinor dyads in obtaining the commutation relations (3.128) is by no means the only way. Indeed, by transvecting through (3.121) by the appropriate null vectors (2.95) one could, with the aid of (3.120), establish (3.128). However, these two approaches are effectively equivalent.

The *Newman–Penrose field equations* can be obtained in a similar way to (3.128). All the necessary information is contained in (3.84) and (3.85), and so we will use them as our starting point.

The commutator in terms of the curvature spinors can be obtained by adding (3.84), multiplied by $\epsilon_{A'B'}$, to (3.85), multiplied by ϵ_{AB}. Thus, on comparing with (3.68), we have

$$(\nabla_a\nabla_b - \nabla_b\nabla_a)\zeta^D = (\epsilon_{A'B'}X_{ABC}{}^D + \epsilon_{AB}\Phi_{A'B'C}{}^D)\zeta^C. \tag{3.129}$$

Transvecting through by (3.122) yields (see (3.124))

$$\begin{aligned}
A^a\nabla_a(B^b\nabla_b\zeta^D) &- (A^a\nabla_a B^b)(\nabla_b\zeta^D)\\
&- B^b\nabla_b(A^a\nabla_a\zeta^D) + (B^b\nabla_b A^a)(\nabla_a\zeta^D)\\
&= A^aB^b(\epsilon_{A'B'}X_{ABC}{}^D) + \epsilon_{AB}\Phi_{A'B'C}{}^D)\zeta^C.
\end{aligned} \tag{3.130}$$

Consider for a moment the left-hand side of (3.130). By replacing ζ^D by $\epsilon_{\mathbf{D}}{}^D$ and transvecting by $\epsilon_D{}^{\mathbf{E}}$ we have

$$\epsilon_D{}^{\mathbf{E}}\nabla_{\mathbf{AA'}}\nabla_{\mathbf{BB'}}\epsilon_{\mathbf{D}}{}^D - \nabla_{\mathbf{AA'}}(\epsilon_{\mathbf{B}}{}^B\epsilon_{\mathbf{B'}}{}^{B'})\epsilon_D{}^{\mathbf{E}}\nabla_{BB'}\epsilon_{\mathbf{D}}{}^D$$
$$-\epsilon_D{}^{\mathbf{E}}\nabla_{\mathbf{BB'}}\nabla_{\mathbf{AA'}}\epsilon_{\mathbf{D}}{}^D + \nabla_{\mathbf{BB'}}(\epsilon_{\mathbf{A}}{}^A\epsilon_{\mathbf{A'}}{}^{A'})\epsilon_D{}^{\mathbf{E}}\nabla_{AA'}\epsilon_{\mathbf{D}}{}^D.$$

The first term can be written, with the help of (3.104), as

$$\epsilon_D{}^{\mathbf{E}}\nabla_{\mathbf{AA'}}(\gamma_{\mathbf{BB'D}}{}^{\mathbf{P}}\epsilon_{\mathbf{P}}{}^{D}) = \nabla_{\mathbf{AA'}}\gamma_{\mathbf{BB'D}}{}^{\mathbf{P}} + \gamma_{\mathbf{BB'D}}{}^{\mathbf{P}}\gamma_{\mathbf{AA'P}}{}^{\mathbf{E}}.$$

The third term results in the same equation, only in this case $\mathbf{B}, \mathbf{B}'$ is swapped with $\mathbf{A}, \mathbf{A}'$. The fourth term can be expressed by

$$\overline{\gamma}_{\mathbf{BB'A'}}{}^{\mathbf{P'}}\gamma_{\mathbf{AP'D}}{}^{E} + \gamma_{\mathbf{BB'A}}{}^{\mathbf{P}}\gamma_{\mathbf{PA'D}}{}^{\mathbf{E}},$$

again with the aid of (3.104). The third term is clearly obtained by swapping $\mathbf{BB}'$ with $\mathbf{AA}'$.

We now substitute (3.58) into the right-hand side of (3.130). Replacing ζ^C by $\epsilon_{\mathbf{D}}{}^{C}$ and transvecting through by $\epsilon_D{}^{\mathbf{E}}$ gives the spinor dyad form. Collating this information then allows us to rewrite (3.130) as

$$\begin{aligned}\nabla_{\mathbf{AA'}}\gamma_{\mathbf{BB'D}}{}^{\mathbf{E}} - \nabla_{\mathbf{BB'}}\gamma_{\mathbf{AA'}D}{}^{\mathbf{E}} = {} & \gamma_{\mathbf{AA'D}}{}^{\mathbf{P}}\gamma_{\mathbf{BB'P}}{}^{\mathbf{E}} - \gamma_{\mathbf{BB'D}}{}^{\mathbf{P}}\gamma_{\mathbf{AA'P}}{}^{\mathbf{E}} \\ & + \gamma_{\mathbf{AA'B}}{}^{\mathbf{P}}\gamma_{\mathbf{PB'D}}{}^{\mathbf{E}} - \gamma_{\mathbf{BB'A}}{}^{\mathbf{P}}\gamma_{\mathbf{PA'D}}{}^{\mathbf{E}} \\ & + \overline{\gamma}_{\mathbf{AA'B}}{}^{\mathbf{P}}\gamma_{\mathbf{BP'D}}{}^{\mathbf{E}} - \overline{\gamma}_{\mathbf{BB'A'}}{}^{\mathbf{P'}}\gamma_{\mathbf{AP'D}}{}^{\mathbf{E}} \\ & + \epsilon^{\mathbf{EP}}(\epsilon_{\mathbf{A'B'}}\Psi_{\mathbf{ABDP}} + \epsilon_{\mathbf{AB}}\Phi_{\mathbf{DPA'B'}}) \\ & + \epsilon_{\mathbf{A'B'}}(\epsilon_{\mathbf{BD}}\epsilon_{\mathbf{A}}{}^{\mathbf{E}} + \epsilon_{\mathbf{B}}{}^{\mathbf{E}}\epsilon_{\mathbf{AD}})\Lambda. \quad (3.131)\end{aligned}$$

Using (2.38), (3.114) and (3.116) we obtain the explicit form of the Newman–Penrose field equations; namely,

$$\begin{aligned}
D\rho - \overline{\delta}\kappa &= (\rho^2 + \sigma\overline{\sigma}) + (\epsilon + \overline{\epsilon})\rho - \overline{\kappa}\tau - \kappa(3\alpha + \overline{\beta} - \pi) + \Phi_{00} \\
D\alpha - \overline{\delta}\epsilon &= (\rho + \overline{\epsilon} - 2\epsilon)\alpha + \beta\overline{\sigma} - \overline{\beta}\epsilon - \kappa\lambda - \overline{\kappa}\gamma + (\epsilon + \rho)\pi + \Phi_{10} \\
D\lambda - \overline{\delta}\pi &= (\rho - 3\epsilon + \overline{\epsilon})\lambda + \overline{\sigma}\mu + (\pi + \alpha - \overline{\beta})\pi - \nu\overline{\kappa} + \Phi_{20} \\
\Delta\mu - \delta\nu &= -(\mu + \gamma + \overline{\gamma})\mu - \lambda\overline{\lambda} + \overline{\nu}\pi + (\overline{\alpha} + 3\beta - \tau)\nu - \Phi_{22} \\
\Delta\beta - \delta\gamma &= (\overline{\alpha} + \beta - \tau)\gamma - \mu\tau + \sigma\nu + \epsilon\overline{\nu} + (\gamma - \overline{\gamma} - \mu)\beta - \alpha\overline{\lambda} - \Phi_{12} \\
\Delta\sigma - \delta\tau &= -(\mu - 3\gamma + \overline{\gamma})\sigma - \overline{\lambda}\rho - (\tau + \beta - \overline{\alpha})\tau + \kappa\overline{\nu} - \Phi_{02} \\
D\sigma - \delta\kappa &= (\rho + \overline{\rho} + 3\epsilon - \overline{\epsilon})\sigma - (\tau - \overline{\pi} + \overline{\alpha} + 3\beta)\kappa + \Psi_0 \\
D\beta - \delta\epsilon &= (\alpha + \pi)\sigma + (\overline{\rho} - \overline{\epsilon})\beta - (\mu + \gamma)\kappa - (\overline{\alpha} - \overline{\pi})\epsilon + \Psi_1 \\
\Delta\alpha - \overline{\delta}\gamma &= (\rho + \epsilon)\nu - (\tau + \beta)\lambda + (\overline{\gamma} - \overline{\mu})\alpha + (\overline{\beta} - \overline{\tau})\gamma - \Psi_3 \\
\Delta\lambda - \overline{\delta}\nu &= -(\mu + \overline{\mu} + 3\gamma - \overline{\gamma})\lambda + (3\alpha + \overline{\beta} + \pi - \overline{\tau})\nu - \Psi_4 \\
D\tau - \Delta\kappa &= (\tau + \overline{\pi})\rho + (\overline{\tau} + \pi)\sigma + (\epsilon - \overline{\epsilon})\tau - (3\gamma + \overline{\gamma})\kappa + \Psi_1 + \Phi_{01} \\
\delta\lambda - \overline{\delta}\mu &= (\rho - \overline{\rho})\nu + (\mu - \overline{\mu})\pi + (\alpha + \overline{\beta})\mu + \overline{\alpha} - 3\beta)\lambda - \Psi_3 + \Phi_{21} \\
D\nu - \Delta\pi &= (\pi + \overline{\tau})\mu + (\overline{\pi} + \tau)\lambda + (\gamma - \overline{\gamma})\pi - (3\epsilon + \overline{\epsilon})\nu + \Psi_3 + \Phi_{21} \\
\delta\rho - \overline{\delta}\sigma &= (\overline{\alpha} + \beta)\rho - (3\alpha - \overline{\beta})\sigma + (\rho - \overline{\rho})\tau + (\mu - \overline{\mu})\kappa - \Psi_1 + \Phi_{01}
\end{aligned}$$

$$\begin{aligned}
D\gamma - \Delta\epsilon &= (\tau + \overline{\pi})\alpha + (\overline{\tau} + \pi)\beta - (\epsilon + \overline{\epsilon})\gamma - (\gamma + \overline{\gamma})\epsilon + \tau\pi - \nu\kappa \\
&\quad + \Psi_2 - \Lambda + \Phi_{11} \\
\delta\alpha - \overline{\delta}\beta &= \mu\rho - \lambda\sigma + \alpha\overline{\alpha} + \beta\overline{\beta} - 2\alpha\beta + (\rho - \overline{\rho})\gamma + (\mu - \overline{\mu})\epsilon \\
&\quad - \Psi_2 + \Lambda + \Phi_{11} \\
D\mu - \delta\pi &= (\overline{\rho} - \epsilon - \overline{\epsilon})\mu + \sigma\lambda + (\overline{\pi} - \overline{\alpha} + \beta)\pi - \nu\kappa + \Psi_2 + 2\Lambda \\
\Delta\rho - \overline{\delta}\tau &= (\gamma + \overline{\gamma} - \overline{\mu})\rho - \sigma\lambda + (\overline{\beta} - \alpha - \overline{\tau})\tau + \nu\kappa - \Psi_2 - 2\Lambda \qquad (3.132)
\end{aligned}$$

where we have used the following simplification in symbolism to represent the dyad components of the Weyl spinor Ψ_{ABCD}:

$$\begin{aligned}
\Psi_0 &= \Psi_{0000} = \Psi_{ABCD} o^A o^B o^C o^D \\
\Psi_1 &= \Psi_{0001} = \Psi_{ABCD} o^A o^B o^C \iota^D \\
\Psi_2 &= \Psi_{0011} = \Psi_{ABCD} o^A o^B \iota^C \iota^D \\
\Psi_3 &= \Psi_{0111} = \Psi_{ABCD} o^A \iota^B \iota^C \iota^D \\
\Psi_4 &= \Psi_{1111} = \Psi_{ABCD} \iota^A \iota^B \iota^C \iota^D, \qquad (3.133)
\end{aligned}$$

and the Ricci spinor $\Phi_{ABC'D'}$:

$$\begin{aligned}
\Phi_{00} &= \Phi_{000'0'} = \Phi_{ABC'D'} o^A o^B o^{C'} o^{D'} \\
\Phi_{10} &= \Phi_{010'0'} = \Phi_{ABC'D'} o^A \iota^B o^{C'} o^{D'} \\
\Phi_{20} &= \Phi_{110'0'} = \Phi_{ABC'D'} \iota^A \iota^B o^{C'} o^{D'} \\
\Phi_{01} &= \Phi_{000'1'} = \Phi_{ABC'D'} o^A o^B o^{C'} \iota^{D'} \\
\Phi_{11} &= \Phi_{010'1'} = \Phi_{ABC'D'} o^A \iota^B o^{C'} \iota^{D'} \\
\Phi_{21} &= \Phi_{110'1'} = \Phi_{ABC'D'} \iota^A \iota^B o^{C'} \iota^{D'} \\
\Phi_{02} &= \Phi_{001'1'} = \Phi_{ABC'D'} o^A o^B \iota^{C'} \iota^{D'} \\
\Phi_{12} &= \Phi_{011'1'} = \Phi_{ABC'D'} o^A \iota^B \iota^{C'} \iota^{D'} \\
\Phi_{22} &= \Phi_{111'1'} = \Phi_{ABC'D'} \iota^A \iota^B \iota^{C'} \iota^{D'}. \qquad (3.134)
\end{aligned}$$

By employing (3.61), the Weyl scalars in (3.132) can be given in terms of the null tetrad vectors defined by (2.95). Thus

$$\begin{aligned}
\Psi_0 &= C_{abcd} l^a m^b l^c m^d = C_{(1)(3)(1)(3)} \\
\Psi_1 &= C_{abcd} l^a m^b l^c n^d = C_{(1)(3)(1)(2)} \\
\Psi_2 &= C_{abcd} l^a m^b \overline{m}^c n^d = C_{(1)(3)(4)(2)} \\
\Psi_3 &= C_{abcd} l^a n^b \overline{m}^c n^d = C_{(1)(2)(4)(3)} \\
\Psi_4 &= C_{abcd} \overline{m}^a n^b \overline{m}^c n^d = C_{(4)(2)(4)(2)}. \qquad (3.135)
\end{aligned}$$

Also, in terms of null tetrad vectors, (3.134) becomes, with the aid of (3.40),

$$\begin{aligned}
\Phi_{00} &= -\frac{1}{2}R_{ab}l^a l^b = -\frac{1}{2}R_{(1)(1)} \\
\Phi_{01} &= -\frac{1}{2}R_{ab}l^a m^b = -\frac{1}{2}R_{(1)(3)} \\
\Phi_{02} &= -\frac{1}{2}R_{ab}m^a m^b = -\frac{1}{2}R_{(3)(3)} \\
\Phi_{10} &= -\frac{1}{2}R_{ab}l^a \overline{m}^b = -\frac{1}{2}R_{(1)(4)} \\
\Phi_{11} &= -\frac{1}{2}R_{ab}l^a n^a + 3\Lambda = -\frac{1}{2}R_{(1)(2)} + \frac{1}{8}R \\
\Phi_{12} &= -\frac{1}{2}R_{ab}m^a n^b = -\frac{1}{2}R_{(3)(2)} \\
\Phi_{20} &= -\frac{1}{2}R_{ab}\overline{m}^a \overline{m}^b = -\frac{1}{2}R_{(4)(4)} \\
\Phi_{21} &= -\frac{1}{2}R_{ab}\overline{m}^a n^b = -\frac{1}{2}R_{(4)(2)} \\
\Phi_{22} &= -\frac{1}{2}R_{ab}n^a n^b = -\frac{1}{2}R_{(2)(2)}.
\end{aligned} \tag{3.136}$$

Note that the bracketed indices in (3.135) and (3.136) are tetrad indices (see Sec. 2.72), Also, (3.136) exhibits all nine independent real components of the Ricci tensor, whereas, of course, (3.135) contains half of the ten independent components of the Weyl tensor. The remaining five are trivially obtained and will be left as an exercise for the reader.

We conclude this section by translating the Bianchi identities (3.95) into spin coefficient form. As the computation is straightforward although slightly tedious (see Exercise 3.10) we quote the result only:

$$\begin{aligned}
&D\Psi_1 - \overline{\delta}\Psi_0 - D\Phi_{01} + \delta\Phi_{00} \\
&\qquad = (\pi - 4\alpha)\Psi_0 + 2(2\rho + \epsilon)\Psi_1 - 3\kappa\Psi_2 \\
&\qquad -(\overline{\pi} - 2\overline{\alpha} - 2\beta)\Phi_{00} - 2(\overline{\rho} + \epsilon)\Phi_{01} \\
&\qquad -2\sigma\Phi_{10} + 2\kappa\Phi_{11} + \overline{\kappa}\Phi_{02} \\
&\Delta\Psi_0 - \delta\Psi_1 + D\Phi_{02} - \delta\Phi_{01} \\
&\qquad = (4\gamma - \mu)\Psi_0 - 2(2\tau + \beta)\Psi_1 + 3\sigma\Psi_2 \\
&\qquad -\overline{\lambda}\Phi_{00} + 2(\overline{\pi} - \beta)\Phi_{01} + 2\sigma\Phi_{11} \\
&\qquad +(\overline{\rho} + 2\epsilon - 2\overline{\epsilon})\Phi_{02} - 2\kappa\Phi_{12} \\
&D\Psi_2 - \overline{\delta}\Psi_1 + \Delta\Phi_{00} - \overline{\delta}\Phi_{01} + 2D\Lambda \\
&\qquad = -\lambda\Psi_0 + 2(\pi - \alpha)\Psi_1 + 3\rho\Psi_2 - 2\kappa\Psi_3
\end{aligned}$$

$$+(2\gamma + 2\overline{\gamma} - \overline{\mu})\Phi_{00} - 2(\alpha + \overline{\tau})\Phi_{01}$$
$$-2\tau\Phi_{10} + 2\rho\Phi_{11} + \overline{\sigma}\Phi_{02}$$

$$\Delta\Psi_1 - \delta\Psi_2 - \Delta\Phi_{01} + \overline{\delta}\Phi_{02} - 2\delta\Lambda$$
$$= \nu\Psi_0 + 2(\gamma - \mu)\Psi_1 - 3\tau\Psi_2 + 2\sigma\Psi_3$$
$$-\overline{\nu}\Phi_{00} + 2(\overline{\mu} - \gamma)\Phi_{01}$$
$$+(2\alpha + \overline{\tau} - 2\overline{\beta})\Phi_{02} + 2\tau\Phi_{11} - 2\rho\Phi_{12}$$

$$D\Psi_3 - \overline{\delta}\Psi_2 - D\Phi_{21} + \delta\Phi_{20} - 2\overline{\delta}\Lambda$$
$$= -2\lambda\Psi_1 + 3\pi\Psi_2 + 2(\rho - \epsilon)\Psi_3 - \kappa\Psi_4$$
$$+2\mu\Phi_{10} - 2\pi\Phi_{11} - (2\beta + \overline{\pi} - 2\overline{\alpha})\Phi_{20}$$
$$-2(\overline{\rho} - \epsilon)\Phi_{21} + \overline{\kappa}\Phi_{22}$$

$$\Delta\Psi_2 - \delta\Psi_3 + D\Phi_{22} - \delta\Phi_{21} + 2\Delta\Lambda$$
$$= 2\nu\Psi_1 - 3\mu\Psi_2 + 2(\beta - \tau)\Psi_3 + \sigma\Psi_4$$
$$-2\mu\Phi_{11} - \overline{\lambda}\Phi_{20} + 2\pi\Phi_{12}$$
$$+2(\beta + \overline{\pi})\Phi_{21} + (\overline{\rho} - 2\epsilon - 2\overline{\epsilon})\Phi_{22}$$

$$D\Psi_4 - \overline{\delta}\Psi_3 + \Delta\Phi_{20} - \overline{\delta}\Phi_{21}$$
$$= -3\lambda\Psi_2 + 2(\alpha + 2\pi)\Psi_3 + (\rho - 4\epsilon)\Psi_4$$
$$+2\nu\Phi_{10} - 2\lambda\Phi_{11} - (2\gamma - 2\overline{\gamma} + \overline{\mu})\Phi_{20}$$
$$-2(\overline{\tau} - \alpha)\Phi_{21} + \overline{\sigma}\Phi_{22}$$

$$\Delta\Psi_3 - \delta\Psi_4 - \Delta\Phi_{21} + \overline{\delta}\Phi_{22}$$
$$= 3\nu\Psi_2 - 2(\gamma + 2\mu)\Psi_3 + (4\beta - \tau)\Psi_4$$
$$-2\nu\Phi_{11} - \overline{\nu}\Phi_{20} + 2\lambda\Phi_{12}$$
$$+2(\gamma + \overline{\mu})\Phi_{21} + (\overline{\tau} - 2\overline{\beta} - 2\alpha)\Phi_{22}$$

$$D\Phi_{11} - \delta\Phi_{10} + \Delta\Phi_{00} - \overline{\delta}\Phi_{01} + 3D\Lambda$$
$$= (2\gamma + 2\overline{\gamma} - \mu - \overline{\mu})\Phi_{00}$$
$$+(\pi - 2\alpha - 2\overline{\tau})\Phi_{01}$$
$$+(\overline{\pi} - 2\overline{\alpha} - 2\tau)\Phi_{10} + 2(\rho + \overline{\rho})\Phi_{11}$$
$$+\overline{\sigma}\Phi_{02} + \sigma\Phi_{20} - \overline{\kappa}\Phi_{12} - \kappa\Phi_{21}$$

$$D\Phi_{12} - \delta\Phi_{11} + \Delta\Phi_{01} - \overline{\delta}\Phi_{02} + 3\delta\Lambda$$
$$= (2\gamma - \mu - 2\overline{\mu})\Phi_{01} + \overline{\nu}\Phi_{00} - \overline{\lambda}\Phi_{10}$$
$$+2(\overline{\pi} - \tau)\Phi_{11}$$
$$+(\pi + 2\overline{\beta} - 2\alpha - \overline{\tau})\Phi_{02}$$
$$+(2\rho + \overline{\rho} - 2\overline{\epsilon})\Phi_{12} + \sigma\Phi_{21} - \kappa\Phi_{22}$$

$$\begin{aligned} D\Phi_{22} - \delta\Phi_{21} + \Delta\Phi_{11} - \overline{\delta}\Phi_{12} + 3\Delta\Lambda \\ = \nu\Phi_{01} + \overline{\nu}\Phi_{10} - 2(\mu + \overline{\mu})\Phi_{11} - \lambda\Phi_{02} \\ -\overline{\lambda}\Phi_{20} + (2\pi - \overline{\tau} + 2\overline{\beta})\Phi_{12} \\ +(2\beta - \tau + 2\overline{\pi})\Phi_{21} \\ +(\rho + \overline{\rho} - 2\epsilon - 2\overline{\epsilon})\Phi_{22}. \qquad (3.137) \end{aligned}$$

Note that to obtain this particular form we have used (3.97) in conjunction with (3.95).

3.6 Newman–Penrose quantities under Lorentz transformations

It may perhaps be prudent to point out at this stage that in fact many calculations involving *Newman–Penrose quantities*, e.g. spin coefficients and Weyl scalars, are simplified due to the nature and *symmetry* of the particular problem. In fact in the majority of these problems one finds that certain spin coefficients equate with each other or vanish entirely. Moreover, a problem will sometimes present itself in a rather complicated manner; but with careful analysis and a judicious choice of transformation of Newman–Penrose quantities, a significant reduction in complexity ensues. We therefore now present a comprehensive account of how the Newman–Penrose quantities behave under the six-parameter group of transformations of the null-tetrad, or spin basis transformations, elements of which preserve the form of the orthonormalisation conditions (2.96), or equivalently (2.94). That is, the proper orthochronous Lorentz group $L(4)$, or equivalently, the three-parameter group of $SL(2,\mathbb{C})$ transformations.

Consider a general linear transformation on the null tetrad system:

$$\begin{aligned} l^a &\mapsto a_1 l^a + a_2 n^a + \overline{c}_1 m^a + c_1 \overline{m}^a \\ n^a &\mapsto a_3 l^a + a_4 n^a + \overline{c}_2 m^a + c_2 \overline{m}^a \\ m^a &\mapsto c_3 l^a + c_4 n^a + c_5 m^a + c_6 \overline{m}^a \\ \overline{m}^a &\mapsto \overline{c}_3 l^a + \overline{c}_4 n^a + \overline{c}_5 \overline{m}^a + \overline{c}_6 m^a, \end{aligned} \qquad (3.138)$$

where a_i are real coefficients and c_i are complex.

Imposing the orthonormalisation conditions (2.96) on the 'old' and 'new' tetrads leads to three rotations and a Lorentz boost. Each of these transformations leaves the orthonormalisation conditions invariant.

3.6.1 *Null rotation with* l *fixed*

In this case the null tetrad transforms as

$$
\begin{aligned}
l^a &\mapsto l^a \\
n^a &\mapsto n^a + \bar{c}m^a + c\overline{m}^a + c\bar{c}l^a \\
m^a &\mapsto m^a + cl^a \\
\overline{m}^a &\mapsto \overline{m}^a + \bar{c}l^a
\end{aligned}
\tag{3.139}
$$

where $c = c_2$.

The spin basis transforms as

$$
\begin{aligned}
o^A &\mapsto o^A \\
\iota^A &\mapsto \iota^A + \bar{c}o^A.
\end{aligned}
\tag{3.140}
$$

The spin coefficients transform as

$$
\begin{aligned}
\nu &\mapsto \nu + c\bar{c}\pi + c\lambda + \bar{c}\mu + \bar{c}^2\tau + \bar{c}^3c\kappa + \bar{c}^2c\rho \\
&\quad + \bar{c}^3\sigma + 2\bar{c}\gamma + 2\bar{c}^2c\epsilon + 2c\bar{c}\alpha + 2\bar{c}^2\beta \\
&\quad + \Delta\bar{c} + c\bar{c}D\bar{c} + c\bar{\delta}\bar{c} + \bar{c}\delta\bar{c} \\
\tau &\mapsto \tau + \bar{c}\sigma + c\rho + c\bar{c}\kappa \\
\gamma &\mapsto \gamma + c\bar{c}\epsilon + c\alpha + \bar{c}\beta + \bar{c}\tau + \bar{c}^2c\kappa + c\bar{c}\rho + \bar{c}^2\sigma \\
\mu &\mapsto \mu + c\pi + \bar{c}^2\sigma + \bar{c}^2c\kappa + 2\bar{c}\beta + 2c\bar{c}\epsilon + \delta\bar{c} + cD\bar{c} \\
\sigma &\mapsto \sigma + c\kappa \\
\beta &\mapsto \beta + c\epsilon + \bar{c}\sigma + c\bar{c}\kappa \\
\lambda &\mapsto \lambda + \bar{c}\pi + 2\bar{c}\alpha + 2\bar{c}^2\epsilon + \bar{c}^2\rho + \bar{c}^3\kappa + \bar{\delta}\bar{c} + \bar{c}D\bar{c} \\
\rho &\mapsto \rho + \bar{c}\kappa \\
\alpha &\mapsto \alpha + \bar{c}\epsilon + \bar{c}\rho + \bar{c}^2\kappa \\
\kappa &\mapsto \kappa \\
\epsilon &\mapsto \epsilon + \bar{c}\kappa \\
\pi &\mapsto \pi + 2\bar{c}\epsilon + \bar{c}^2\kappa + D\bar{c}.
\end{aligned}
\tag{3.141}
$$

The components of $\Phi_{ABC'D'}$ transform as

$$
\begin{aligned}
\Phi_{22} &\mapsto \Phi_{22} + 4c\bar{c}\Phi_{11} + 2c\Phi_{21} + 2\bar{c}\Phi_{12} \\
&\quad + \bar{c}^2c^2\Phi_{00} + 2\bar{c}c^2\Phi_{10} + 2\bar{c}^2c\Phi_{01} + \bar{c}^2\Phi_{02} + c^2\Phi_{20} \\
\Phi_{11} &\mapsto \Phi_{11} + c\bar{c}\Phi_{00} + c\Phi_{10} + \bar{c}\Phi_{01} \\
\Phi_{21} &\mapsto \Phi_{21} + 2c\bar{c}\Phi_{10} + \bar{c}^2\Phi_{01} + c\Phi_{20} + 2\bar{c}\Phi_{11} + \bar{c}^2c\Phi_{00}
\end{aligned}
$$

$$
\begin{aligned}
\Phi_{10} &\mapsto \Phi_{10} + \bar{c}\Phi_{00} \\
\Phi_{12} &\mapsto \Phi_{12} + 2c\bar{c}\Phi_{01} + c^2\Phi_{10} + \bar{c}\Phi_{02} + 2c\Phi_{11} + c^2\bar{c}\Phi_{00} \\
\Phi_{01} &\mapsto \Phi_{01} + c\Phi_{00} \\
\Phi_{20} &\mapsto \Phi_{20} + 2\bar{c}\Phi_{10} + \bar{c}^2\Phi_{00} \\
\Phi_{00} &\mapsto \Phi_{00} \\
\Phi_{02} &\mapsto \Phi_{02} + 2c\Phi_{01} + c^2\Phi_{00}.
\end{aligned} \tag{3.142}
$$

Also, $\Lambda \mapsto \Lambda$.

The components of Ψ_{ABCD} transform as

$$
\begin{aligned}
\Psi_0 &\mapsto \Psi_0 \\
\Psi_1 &\mapsto \Psi_1 + \bar{c}\Psi_0 \\
\Psi_2 &\mapsto \Psi_2 + 2\bar{c}\Psi_1 + \bar{c}^2\Psi_0 \\
\Psi_3 &\mapsto \Psi_3 + 3\bar{c}\Psi_2 + 3\bar{c}^2\Psi_1 + \bar{c}^3\Psi_0 \\
\Psi_4 &\mapsto \Psi_4 + 4\bar{c}\Psi_3 + 6\bar{c}^2\Psi_2 + 4\bar{c}^3\Psi_1 + \bar{c}^4\Psi_0.
\end{aligned} \tag{3.143}
$$

3.6.2 *Null rotation with* n *fixed*

The null tetrad transforms as

$$
\begin{aligned}
l^a &\mapsto l^a + \bar{c}m^a + c\overline{m}^a + c\bar{c}n^a \\
n^a &\mapsto n^a \\
m^a &\mapsto m^a + cn^a \\
\overline{m}^a &\mapsto \overline{m}^a + \bar{c}n^a
\end{aligned} \tag{3.144}
$$

where $c = c_1$.

The spin basis transforms as

$$
\begin{aligned}
o^A &\mapsto o^A + c\iota^A \\
\iota^A &\mapsto \iota^A.
\end{aligned} \tag{3.145}
$$

The spin coefficients transform as

$$
\begin{aligned}
\kappa &\mapsto \kappa + c\bar{c}\tau + \bar{c}\sigma + c\rho + c^2\pi + c^3\bar{c}\nu + c^2\bar{c}\mu + c^3\lambda + c\epsilon + 2c^2\bar{c}\gamma \\
&\quad +2c\bar{c}\beta + 2c^2\alpha - Dc - c\bar{c}\Delta c - \bar{c}\delta c - c\bar{\delta}c \\
\pi &\mapsto \pi + c\lambda + \bar{c}\mu + c\bar{c}\nu \\
\epsilon &\mapsto \epsilon + c\bar{c}\gamma + \bar{c}\beta + c\alpha + c\pi + c^2\bar{c}\nu + c\bar{c}\mu + c^2\lambda \\
\rho &\mapsto \rho + \bar{c}\tau + c^2\lambda + c^2\bar{c}\nu + 2c\alpha + 2c\bar{c}\gamma - \bar{\delta}c - \bar{c}\Delta c \\
\lambda &\mapsto \lambda + \bar{c}\nu
\end{aligned}
$$

$$
\begin{aligned}
&\alpha \mapsto \alpha + \bar{c}\tau + c\lambda + c\bar{c}\nu \\
&\sigma \mapsto \sigma + c\tau + 2c\beta + 2c^2\gamma + c^2\mu + c^3\nu - \delta c - c\Delta c \\
&\mu \mapsto \mu + c\nu \\
&\beta \mapsto \beta + c\gamma + c\mu + c^2\nu \\
&\nu \mapsto \nu \\
&\gamma \mapsto \gamma + c\nu \\
&\tau \mapsto \tau + 2c\gamma + c^2\nu - \Delta c.
\end{aligned} \tag{3.146}
$$

The components of $\Phi_{ABC'D'}$ transform as

$$
\begin{aligned}
\Phi_{00} \mapsto\ & \Phi_{00} + 4c\bar{c}\Phi_{11} + 2\bar{c}\Phi_{01} + 2c\Phi_{10} + c^2\bar{c}^2\Phi_{22} \\
& +2c\bar{c}^2\Phi_{12} + 2c^2\bar{c}\Phi_{21} + c^2\Phi_{20} + \bar{c}^2\Phi_{02} \\
\Phi_{11} \mapsto\ & \Phi_{11} + c\bar{c}\Phi_{22} + \bar{c}\Phi_{12} + c\Phi_{21} \\
\Phi_{01} \mapsto\ & \Phi_{01} + 2c\bar{c}\Phi_{12} + c^2\Phi_{21} + \bar{c}\Phi_{02} + 2c\Phi_{11} + c^2\bar{c}\Phi_{22} \\
\Phi_{12} \mapsto\ & \Phi_{12} + c\Phi_{22} \\
\Phi_{10} \mapsto\ & \Phi_{10} + 2c\bar{c}\Phi_{21} + \bar{c}\Phi_{12} + c\Phi_{20} + 2\bar{c}\Phi_{11} + \bar{c}^2c\Phi_{22} \\
\Phi_{21} \mapsto\ & \Phi_{21} + \bar{c}\Phi_{22} \\
\Phi_{02} \mapsto\ & \Phi_{02} + 2c\Phi_{12} + c^2_{22} \\
\Phi_{22} \mapsto\ & \Phi_{22} \\
\Phi_{20} \mapsto\ & \Phi_{20} + 2\bar{c}\Phi_{21} + \bar{c}^2\Phi_{22}.
\end{aligned} \tag{3.147}
$$

Also, $\Lambda \mapsto \Lambda$.

The components of Ψ_{ABCD} transform as

$$
\begin{aligned}
&\Psi_0 \mapsto \Psi_0 + 4c\Psi_1 + 6c^2\Psi_2 + 4c^3\Psi_3 + c^4\Psi_4 \\
&\Psi_1 \mapsto \Psi_1 + 3c\Psi_2 + 3c^2\Psi_3 + c^3\Psi_4 \\
&\Psi_2 \mapsto \Psi_2 + 2c\Psi_3 + c^2\Psi_4 \\
&\Psi_3 \mapsto \Psi_3 + c\Psi_4 \\
&\Psi_4 \mapsto \Psi_4.
\end{aligned} \tag{3.148}
$$

3.6.3 *Lorentz spin-boost transformation*

A *Lorentz boost* is defined by the following tetrad transformation:

$$
\begin{aligned}
l^a &\mapsto al^a \\
n^a &\mapsto \frac{1}{a}n^a \\
m^a &\mapsto m^a
\end{aligned}
$$

$$\overline{m}^a \mapsto \overline{m}^a \tag{3.149}$$

where $a = a_1 = 1/a_4$. We call a the *boost parameter*, which is, of course, real.

A *spin transformation* (or *spatial rotation*) is defined by the following tetrad transformation

$$\begin{aligned} l^a &\mapsto l^a \\ n^a &\mapsto n^a \\ m^a &\mapsto e^{i\theta} m^a \\ \overline{m}^a &\mapsto e^{-i\theta}\overline{m}^a \end{aligned} \tag{3.150}$$

where $e^{i\theta} = c_5$. We call θ the *spin parameter* (or *spatial rotation parameter*), which again must be real.

The null tetrad of the combined spin-boost transformation transforms as

$$\begin{aligned} l^a &\mapsto a l^a \\ n^a &\mapsto \frac{1}{a} n^a \\ m^a &\mapsto e^{i\theta} m^a \\ \overline{m}^a &\mapsto e^{-i\theta}\overline{m}^a. \end{aligned} \tag{3.151}$$

The spin basis transforms as

$$\begin{aligned} o^A &\mapsto a^{1/2} e^{i\theta/2} o^A \\ \iota^A &\mapsto a^{-1/2} e^{-i\theta/2} \iota^A. \end{aligned} \tag{3.152}$$

The spin coefficients transform as

$$\begin{aligned} &\kappa \mapsto a^2 e^{i\theta}\kappa \\ &\pi \mapsto e^{-i\theta}\pi \\ &\epsilon \mapsto \frac{a}{2}\left[\frac{1}{a} Da + e^{-i\theta} De^{i\theta} + 2\epsilon\right] \\ &\rho \mapsto a\rho \\ &\lambda \mapsto \frac{1}{a} e^{-2i\theta}\lambda \\ &\alpha \mapsto \frac{1}{2} e^{-i\theta}\left[\frac{1}{a}\overline{\delta} a + e^{-i\theta}\overline{\delta} e^{i\theta} + 2\alpha\right] \\ &\sigma \mapsto a e^{2i\theta}\sigma \\ &\mu \mapsto \frac{1}{a}\mu \end{aligned}$$

$$\beta \mapsto \frac{1}{2}e^{-i\theta}\left[\frac{1}{a}\delta a + e^{-i\theta}\delta e^{i\theta} + 2\beta\right]$$
$$\nu \mapsto \frac{1}{a^2}e^{-i\theta}\nu$$
$$\gamma \mapsto \frac{1}{2a}\left[\frac{1}{a}\Delta a + e^{-i\theta}\Delta e^{i\theta} + 2\gamma\right]$$
$$\tau \mapsto e^{i\theta}\tau. \tag{3.153}$$

The components of $\Phi_{ABC'D'}$ transform as

$$\Phi_{00} \mapsto a^2\Phi_{00}$$
$$\Phi_{11} \mapsto \Phi_{11}$$
$$\Phi_{01} \mapsto ae^{i\theta}\Phi_{01}$$
$$\Phi_{12} \mapsto a^{-1}e^{i\theta}\Phi_{12}$$
$$\Phi_{10} \mapsto ae^{-i\theta}\Phi_{10}$$
$$\Phi_{21} \mapsto a^{-1}e^{-i\theta}\Phi_{21}$$
$$\Phi_{02} \mapsto e^{2i\theta}\Phi_{02}$$
$$\Phi_{22} \mapsto a^{-2}\Phi_{22}$$
$$\Phi_{20} \mapsto e^{-2i\theta}\Phi_{20}. \tag{3.154}$$

Also, $\Lambda \mapsto \Lambda$.

The components of Ψ_{ABCD} transform as

$$\Psi_0 \mapsto a^2e^{2i\theta}\Psi_0$$
$$\Psi_1 \mapsto ae^{i\theta}\Psi_1$$
$$\Psi_2 \mapsto \Psi_2$$
$$\Psi_3 \mapsto a^{-1}e^{-i\theta}\Psi_3$$
$$\Psi_4 \mapsto a^{-2}e^{-2i\theta}\Psi_4. \tag{3.155}$$

3.7 Miscellaneous transformations

It is sometimes convenient to make the following transformations on the null tetrad

3.7.1 $\mathbf{l} \leftrightarrow \mathbf{n}$ *and* $\mathbf{m} \leftrightarrow -\overline{\mathbf{m}}$

$$l^a \mapsto n^a$$
$$n^a \mapsto l^a$$

$$\begin{aligned} m^a &\mapsto -\overline{m}^a \\ \overline{m}^a &\mapsto -m^a. \end{aligned} \tag{3.156}$$

The spin basis transforms as

$$\begin{aligned} o^A &\mapsto \iota^A \\ \iota^A &\mapsto -o^A. \end{aligned} \tag{3.157}$$

The spin coefficients transform as

$$\begin{aligned} \rho &\mapsto -\mu \\ \alpha &\mapsto \beta \\ \lambda &\mapsto -\sigma \\ \kappa &\mapsto \nu \\ \epsilon &\mapsto -\gamma \\ \pi &\mapsto \tau \\ \sigma &\mapsto -\lambda \\ \beta &\mapsto \alpha \\ \mu &\mapsto -\rho \\ \tau &\mapsto \pi \\ \gamma &\mapsto -\epsilon \\ \nu &\mapsto \kappa. \end{aligned} \tag{3.158}$$

The components of $\Phi_{ABC'D'}$ transform as

$$\begin{aligned} \Phi_{00} &\mapsto \Phi_{22} \\ \Phi_{11} &\mapsto \Phi_{11} \\ \Phi_{22} &\mapsto \Phi_{00} \\ \Phi_{10} &\mapsto -\Phi_{12} \\ \Phi_{20} &\mapsto \Phi_{02} \\ \Phi_{12} &\mapsto -\Phi_{10} \\ \Phi_{01} &\mapsto -\Phi_{21} \\ \Phi_{02} &\mapsto \Phi_{20} \\ \Phi_{21} &\mapsto -\Phi_{01}. \end{aligned} \tag{3.159}$$

Also, $\Lambda \mapsto \Lambda$.

The components of Ψ_{ABCD} transform as

$$\begin{aligned}\Psi_0 &\mapsto \Psi_4\\ \Psi_1 &\mapsto -\Psi_3\\ \Psi_2 &\mapsto \Psi_2\\ \Psi_3 &\mapsto -\Psi_1\\ \Psi_4 &\mapsto \Psi_0.\end{aligned} \tag{3.160}$$

There are two other transformations worth mentioning which can be used to make great simplifications. Moreover, if an equation has been yielded by an alternative process these transformations can function as a test for checking possible errors incurred during calculations.

3.7.2 *Prime operation*

The null tetrad transforms as

$$\begin{aligned}l^a &\mapsto n^a\\ n^a &\mapsto l^a\\ m^a &\mapsto \overline{m}^a\\ \overline{m}^a &\mapsto m^a.\end{aligned} \tag{3.161}$$

The spin basis transforms as

$$\begin{aligned}o^A &\mapsto i\iota^A\\ \iota^A &\mapsto io^A\\ o^{A'} &\mapsto -i\iota^{A'}\\ \iota^{A'} &\mapsto -io^{A'}.\end{aligned} \tag{3.162}$$

Or, using the prime notation,

$$\begin{aligned}(l^a)' &= n^a\\ (n^a)' &= l^a\\ (m^a)' &= \overline{m}^a\\ (\overline{m}^a)' &= m^a.\end{aligned} \tag{3.163}$$

and

$$\begin{aligned}(o^A)' &= i\iota^A\\ (\iota^A)' &= io^A\end{aligned}$$

$$(o^{A'})' = -i\iota^{A'}$$
$$(\iota^{A'})' = -io^{A'}. \tag{3.164}$$

The spin coefficients transform as

$$\kappa' = -\nu$$
$$\pi' = -\tau$$
$$\epsilon' = -\gamma$$
$$\rho' = -\mu$$
$$\lambda' = -\sigma$$
$$\alpha' = -\beta$$
$$\sigma' = -\lambda$$
$$\mu' = -\rho$$
$$\beta' = -\alpha$$
$$\nu' = -\kappa$$
$$\gamma' = -\epsilon$$
$$\tau' = -\pi. \tag{3.165}$$

The components of $\Phi_{ABC'D'}$ transform as

$$\Phi'_{00} = \Phi_{22}$$
$$\Phi'_{11} = \Phi_{11}$$
$$\Phi'_{22} = \Phi_{00}$$
$$\Phi'_{10} = \Phi_{12}$$
$$\Phi'_{20} = \Phi_{02}$$
$$\Phi'_{12} = \Phi_{10}$$
$$\Phi'_{01} = \Phi_{21}$$
$$\Phi'_{02} = \Phi_{20}$$
$$\Phi'_{21} = \Phi_{01}. \tag{3.166}$$

Also, $\Lambda' = \Lambda$.

The components of Ψ_{ABCD} transform as

$$\Psi'_0 = \Psi_4$$
$$\Psi'_1 = \Psi_3$$
$$\Psi'_2 = \Psi_2$$
$$\Psi'_3 = \Psi_1$$

$$\Psi'_4 = \Psi_0. \tag{3.167}$$

3.7.3 *Asterisk operation*

The null tetrad transforms as

$$\begin{aligned} l^a &\mapsto m^a \\ n^a &\mapsto -\overline{m}^a \\ m^a &\mapsto -l^a \\ \overline{m}^a &\mapsto n^a. \end{aligned} \tag{3.168}$$

The spin basis transforms as

$$\begin{aligned} o^A &\mapsto o^A \\ \iota^A &\mapsto \iota^A \\ \overline{o}^{A'} &\mapsto -\overline{\iota}^{A'} \\ \overline{\iota}^{A'} &\mapsto -\overline{o}^{A'}. \end{aligned} \tag{3.169}$$

Or, using the asterisk notation,

$$\begin{aligned} (l^a)^* &= m^a \\ (n^a)^* &= -\overline{m}^a \\ (m^a)^* &= -l^a \\ (\overline{m}^a)^* &= n^a \end{aligned} \tag{3.170}$$

and

$$\begin{aligned} (o^A)^* &= o^A \\ (\iota^A)^* &= \iota^A \\ (o^{A'})^* &= \iota^{A'} \\ (\iota^{A'})^* &= -o^{A'}. \end{aligned} \tag{3.171}$$

The spin coefficients transform as

$$\begin{aligned} \kappa^* &= \sigma \\ \pi^* &= \mu \\ \epsilon^* &= \beta \\ \rho^* &= \tau \\ \lambda^* &= \nu \\ \alpha^* &= \gamma \end{aligned}$$

$$\begin{aligned}
\sigma^* &= -\kappa \\
\mu^* &= -\pi \\
\beta^* &= -\epsilon \\
\nu^* &= -\lambda \\
\gamma^* &= -\alpha \\
\tau^* &= -\rho,
\end{aligned} \tag{3.172}$$

and

$$\begin{aligned}
(\overline{\kappa})^* &= \overline{\lambda} \\
(\overline{\pi})^* &= \overline{\rho} \\
(\overline{\epsilon})^* &= -\overline{\alpha} \\
(\overline{\rho})^* &= -\overline{\pi} \\
(\overline{\lambda})^* &= -\overline{\kappa} \\
(\overline{\alpha})^* &= \overline{\epsilon} \\
(\overline{\sigma})^* &= \overline{\nu} \\
(\overline{\mu})^* &= \overline{\tau} \\
(\overline{\beta})^* &= -\overline{\gamma} \\
(\overline{\nu})^* &= -\overline{\sigma} \\
(\overline{\gamma})^* &= \overline{\beta} \\
(\overline{\tau})^* &= -\overline{\mu}.
\end{aligned} \tag{3.173}$$

The components of $\Phi_{ABC'D'}$ transform as

$$\begin{aligned}
\Phi^*_{00} &= \Phi_{02} \\
\Phi^*_{11} &= -\Phi_{11} \\
\Phi^*_{22} &= \Phi_{20} \\
\Phi^*_{10} &= \Phi_{12} \\
\Phi^*_{20} &= \Phi_{22} \\
\Phi^*_{12} &= \Phi_{10} \\
\Phi^*_{01} &= -\Phi_{01} \\
\Phi^*_{02} &= \Phi_{00} \\
\Phi^*_{21} &= -\Phi_{21}.
\end{aligned} \tag{3.174}$$

Also, $\Lambda^* = \Lambda$.

The components of Ψ_{ABCD} transform as

$$\begin{aligned}
\Psi_0^* &= \Psi_0 \\
\Psi_1^* &= \Psi_1 \\
\Psi_2^* &= \Psi_2 \\
\Psi_3^* &= \Psi_3 \\
\Psi_4^* &= \Psi_4,
\end{aligned} \tag{3.175}$$

and

$$\begin{aligned}
\overline{\Psi}_0^* &= \overline{\Psi}_4 \\
\overline{\Psi}_1^* &= -\overline{\Psi}_3 \\
\overline{\Psi}_2^* &= \overline{\Psi}_2 \\
\overline{\Psi}_3^* &= -\overline{\Psi}_1 \\
\overline{\Psi}_4^* &= \overline{\Psi}_0.
\end{aligned} \tag{3.176}$$

Note that (see (3.113)) the asterisk operation (discovered by Sachs) and complex conjugation are non-commutative; however, the prime operation does commute with complex conjugation.

3.8 Geroch–Held–Penrose formalism

There are a number of problems in general relativity where each point of space-time has some structure associated with it in a natural way. That is, one or more null directions may be defined naturally at each point.

Of some interest is the gravitational radiation problem. (Gravitational disturbances propagating at the speed of light were shown to be solutions of the linearised Einstein field equations, (Einstein, 1916). These wavelike solutions were, for many years, treated with suspicion and skepticism — their physical reality being questionable. Whether the mass of a body would shrink while gravitational waves were being transmitted was also debatable until proof was finally obtained in 1962 by Bondi, Sachs et al in favour of mass reduction during radiation emission. However, physical evidence of gravitational waves remained elusive for some time due to the nature of gravitational interactions being particularly weak, and eventually came by virtue of the Hulse–Taylor binary pulsar, (Hulse and Taylor, 1975)). In this particular case, the conditions of the problem usually single out two null directions at each point, which we will see perfectly compliments the case of the Geroch–Held–Penrose formalism (Geroch, Held and Penrose, 1973) (usually abbreviated to GHP formalism).

3.8.1 *Spin- and boost-weighted scalars*

Consider two future-pointing null directions at each point of space-time. Let the null vectors l^a and n^a be the chosen vectors acting in these directions such that the normalisation condition $l^a n_a = 1$ holds. Then the spacelike vectors x^a and y^a (orthogonal to l^a, n^a as well as each other) defined in (2.109) can constitute the two other vectors forming the tetrad. The fact that two null vectors have been singled out means that there is left at each point a two-dimensional 'gauge' freedom. The two null directions are preserved by the two-parameter subgroup of the Lorentz group. The spin-boost transformations (3.151):

$$\begin{aligned} l^a &\mapsto a l^a \\ n^a &\mapsto \frac{1}{a} n^a \\ m^a &\mapsto e^{i\theta} m^a \\ \overline{m}^a &\mapsto e^{-i\theta} \overline{m}^a, \end{aligned} \tag{3.177}$$

or (3.152):

$$\begin{aligned} o^A &\mapsto a^{1/2} e^{i\theta/2} o^A \\ \iota^A &\mapsto a^{-1/2} e^{-i\theta/2} \iota^A \end{aligned} \tag{3.178}$$

generate this group and leave the null directions l^a and n^a invariant. The spinor fields o^A and ι^A obey the usual normalisation condition

$$o_A \iota^A = 1 \tag{3.179}$$

on the space-time.

It will be convenient to replace the boost parameter, a, and the spin parameter, θ, by a single complex number λ (not to be confused with the spin coefficient λ) such that

$$\lambda^2 = a e^{i\theta}. \tag{3.180}$$

The gauge group at each point will then constitute the multiplicative group of complex numbers λ. Using (3.180) we find that (3.177) and (3.178) transform as follows:

$$
\begin{aligned}
l^a &\mapsto \lambda\overline{\lambda} l^a \\
n^a &\mapsto \lambda^{-1}\overline{\lambda}^{-1} n^a \\
m^a &\mapsto \lambda\overline{\lambda}^{-1} m^a \\
\overline{m}^a &\mapsto \lambda^{-1}\overline{\lambda}\overline{m}^a
\end{aligned} \tag{3.181}
$$

and

$$
\begin{aligned}
o^A &\mapsto \lambda o^A \\
\iota^A &\mapsto \lambda^{-1}\iota^A.
\end{aligned} \tag{3.182}
$$

The GHP formalism is concerned with scalar quantities undergoing a transformation

$$\eta \mapsto \lambda^p \overline{\lambda}^q \eta \tag{3.183}$$

whenever the null tetrad $(l^a, n^a, m^a, \overline{m}^a)$ transforms according to (3.181) or, equivalently, whenever the spin basis transforms according to (3.182). Originally, η was referred to as a *spin- and boost-weighted scalar of type* $\{p, q\}$; it is now more common to discuss η as a *weighted scalar of type* $\{p, q\}$ which has as *spin-weight*

$$s = \frac{1}{2}(p - q) \tag{3.184}$$

and a *boost-weight*

$$r = \frac{1}{2}(p + q), \tag{3.185}$$

or simply as a $\{p, q\}$-scalar.

3.8.2 *Weighted and unweighted spin coefficients*

For a scalar quantity to be a weighted quantity, it must transform homogeneously under the spin-boost transformations of the null tetrad (3.181) or spin basis (3.182) as η does in (3.183), otherwise the scalar quantity is *unweighted.* We will find that not all of the spin coefficients defined in (3.115) transform homogeneously, and so two subdivisions of spin coefficients will be recognised. Furthermore, a slight modification on the labelling of spin coefficients will be introduced. The purpose of this minimal departure in notation is so as to emphasise the close correspondence between pairs of spin coefficients, and is achieved by applying the prime operation of Sec. 3.7.2

to appropriate spin coefficient relations. Thence, the modified form of the spin coefficients is given by

$$
\begin{aligned}
\kappa &= m^a D l_a = o^A D o_A \\
\rho &= m^a \delta' l_a = o^A \delta' o_A \\
\sigma &= m^a \delta l_a = o^A \delta o_A \\
\tau &= m^a D' l_a = o^A D' o_A \\
\kappa' &= \overline{m}^a D' n_a = -\iota^A D' \iota_A \\
\rho' &= \overline{m}^a \delta n_a = -\iota^A \delta \iota_A \\
\sigma' &= \overline{m}^a \delta' n_a = -\iota^A \delta' \iota_A \\
\tau' &= \overline{m}^a D n_a = -\iota^A D \iota_A \\
\beta &= \frac{1}{2}(n^a \delta l_a - \overline{m}^a \delta m_a) = \iota^A \delta o_A \\
\epsilon &= \frac{1}{2}(n^a D l_a - \overline{m}^a D m_a) = \iota^A D o_A \\
\beta' &= \frac{1}{2}(l^a \delta' n_a - m^a \delta' \overline{m}_a) = -o^A \delta' \iota_A \\
\epsilon' &= \frac{1}{2}(l^a D' n_a - m^a D' \overline{m}_a) = -o^A D' \iota_A
\end{aligned} \tag{3.186}
$$

where

$$
\begin{aligned}
D &= \nabla_{00'} = o^A o^{A'} \nabla_{AA'} = l^a \nabla_a \\
D' &= \Delta = \nabla_{11'} = \iota^A \iota^{A'} \nabla_{AA'} = n^a \nabla_a \\
\delta &= \nabla_{01'} = o^A \iota^{A'} \nabla_{AA'} = m^a \nabla_a \\
\delta' &= \overline{\delta} = \nabla_{10'} = \iota^A o^{A'} \nabla_{AA'} = \overline{m}^a \nabla_a .
\end{aligned} \tag{3.187}
$$

(See (1.116)).

The weighted spin coefficients in (3.186) transform as

$$
\begin{aligned}
\kappa &\mapsto \lambda^3 \overline{\lambda} \kappa \\
\kappa' &\mapsto \lambda^{-3} \overline{\lambda}^{-1} \kappa' \\
\rho &\mapsto \lambda \overline{\lambda} \rho \\
\rho' &\mapsto \lambda^{-1} \overline{\lambda}^{-1} \rho \\
\sigma &\mapsto \lambda^3 \overline{\lambda}^{-1} \sigma \\
\sigma' &\mapsto \lambda^{-3} \overline{\lambda} \sigma \\
\tau &\mapsto \lambda \overline{\lambda}^{-1} \tau \\
\tau' &\mapsto \lambda^{-1} \overline{\lambda} \tau' .
\end{aligned} \tag{3.188}
$$

The unweighted spin coefficients in (3.186) transform as

$$\begin{aligned}
\beta &\mapsto \lambda\overline{\lambda}^{-1}\beta + \overline{\lambda}^{-1}\delta\lambda \\
\beta' &\mapsto \lambda^{-1}\overline{\lambda}\beta' - \lambda^{-2}\overline{\lambda}\delta'\lambda \\
\epsilon &\mapsto \lambda\overline{\lambda}\epsilon + \overline{\lambda}D\lambda \\
\epsilon' &\mapsto \lambda^{-1}\overline{\lambda}^{-1}\epsilon' - \lambda^{-2}\overline{\lambda}^{-1}D'\lambda.
\end{aligned} \tag{3.189}$$

Clearly, no type can be assigned to the unweighted spin coefficients in (3.189). The types of the weighted spin coefficients are given by

$$\begin{aligned}
\kappa &: \{3,1\}\text{-scalar} \\
\kappa' &: \{-3,-1\}\text{-scalar} \\
\rho &: \{1,1\}\text{-scalar} \\
\rho' &: \{-1,-1\}\text{-scalar} \\
\sigma &: \{3,-1\}\text{-scalar} \\
\sigma' &: \{-3,1\}\text{-scalar} \\
\tau &: \{1,-1\}\text{-scalar} \\
\tau' &: \{-1,1\}\text{-scalar.}
\end{aligned} \tag{3.190}$$

The scalar quantities expressed in (3.133) and (3.134) are weighted. Under (3.182) they transform according to (3.155) and (3.154) respectively. Then with the aid of (3.180), the types of these weighted scalars are given by

$$\begin{aligned}
\Psi_0 &: \{4,0\}\text{-scalar} \\
\Psi_1 &: \{2,0\}\text{-scalar} \\
\Psi_2 &: \{0,0\}\text{-scalar} \\
\Psi_3 &: \{-2,0\}\text{-scalar} \\
\Psi_4 &: \{-4,0\}\text{-scalar}
\end{aligned} \tag{3.191}$$

and

$$\begin{aligned}
\Phi_{00} &: \{2,2\}\text{-scalar} \\
\Phi_{01} &: \{2,0\}\text{-scalar} \\
\Phi_{02} &: \{2,-2\}\text{-scalar} \\
\Phi_{10} &: \{0,2\}\text{-scalar} \\
\Phi_{11} &: \{0,0\}\text{-scalar} \\
\Phi_{12} &: \{0,-2\}\text{-scalar}
\end{aligned}$$

$$\begin{aligned}\Phi_{20} &: \{-2, 2\}\text{-scalar}\\ \Phi_{21} &: \{-2, 0\}\text{-scalar}\\ \Phi_{22} &: \{-2, -2\}\text{-scalar}.\end{aligned} \tag{3.192}$$

Also,

$$\Lambda : \{0, 0\}\text{-scalar}. \tag{3.193}$$

Any tensor or spinor can be defined by pairing together the tensor field or spinor field, on the space-time, with their associated weighted scalars of different type $\{p, q\}$. Obtaining these from either the tensor or spinor requires contraction with appropriate combinations of null vectors $l^a, n^a, m^a, \overline{m}^a$ or transvection with appropriate combinations of basis spinors $o^A, \iota^A, o^{A'}, \iota^{A'}$, respectively. It follows from (2.95) that exactly identical weighted scalars will be yielded by the fact that tensor fields can be redefined as spinor fields. This is because the null tetrad can be directly interpreted in terms of the spin basis.

The product of two weighted scalars of different type can be found in a straightforward way:

$$[\{p, q\}\text{-scalar}][\{u, v\}\text{-scalar}] = \{p + u, q + v\}\text{-scalar}.$$

The sum of two weighted scalars can be found provided that they are of the same type.

3.8.3 *Weighted differential operators*

The intrinsic derivatives D, D', δ and δ' defined in (3.116) are unweighted operators. That is, they do not produce weighted scalars when they act on a scalar of type $\{p, q\}$, with p and q not both zero. However, this can be remedied by introducing two new differential operators: ð (pronounced 'eth' or 'edth') and þ (pronounced 'thorn'). They are defined in such a way that new differential operators are expressed as combinations of unweighted intrinsic derivatives and unweighted spin coefficients. By doing this the unweighted quantities no longer participate in the formalism. Thus for a $\{p, q\}$-scalar η we define the following form of ð and þ:

$$\begin{aligned}\text{þ}\eta &= (D - p\epsilon - q\overline{\epsilon})\eta\\ \text{þ}'\eta &= (D' + p\epsilon' + q\overline{\epsilon}')\eta\\ \eth\eta &= (\delta - p\beta + q\overline{\beta}')\eta\\ \eth'\eta &= (\delta' + p\beta' + q\overline{\beta})\eta.\end{aligned} \tag{3.194}$$

Note that these operators need not be restricted to acting on scalars, and that it is perfectly permissible for ð and þ to act on tensors and spinors as well. Also, when acting on products they obey Liebniz's law. Because complex conjugation and the prime operation commute with each other, it is irrelevant in which order they are carried out when combined.

With respect to the null tetrad (3.181) or spin basis (3.182), the spin relations in (3.194) transform as

$$
\begin{aligned}
þ\eta &\mapsto (D - p\epsilon - q\bar{\epsilon})\lambda^{(p+1)}\overline{\lambda}^{(q+1)}\eta \\
þ'\eta &\mapsto (D' + p\epsilon' + q\bar{\epsilon}')\lambda^{(p-1)}\overline{\lambda}^{(q-1)}\eta \\
ð\eta &\mapsto (\delta - p\beta + q\overline{\beta}')\lambda^{(p+1)}\overline{\lambda}^{(q-1)}\eta \\
ð'\eta &\mapsto (\delta' + p\beta' - q\overline{\beta})\lambda^{(p-1)}\overline{\lambda}^{(q+1)}\eta.
\end{aligned} \tag{3.195}
$$

When discussing the type there are two possible ways of defining these differential operators. Firstly, if ð or þ acts on a scalar of type $\{u, v\}$ yielding a $\{p+u, q+v\}$-scalar, then we say ð, or þ, has type $\{p, q\}$. Secondly, we may define the type $\{0, 0\}$-operator $\Theta_{AA'}$ by

$$\Theta_{AA'} = \nabla_{AA'} - p\iota^B\nabla_{AA'}o_B - q\iota^{B'}\nabla_{AA'}o_{B'} \tag{3.196}$$

which acts on a quantity of type $\{p, q\}$. Then ð and þ are given by

$$
\begin{aligned}
þ &= \Theta_{00'} \\
þ' &= \Theta_{11'} \\
ð &= \Theta_{01'} \\
ð' &= \Theta_{10'}.
\end{aligned} \tag{3.197}
$$

Equivalently, we can define

$$
\begin{aligned}
\Theta_a &= l_a þ' + n_a þ - m_a ð' - \overline{m}_a ð \\
&= \nabla_a - rn^b\nabla_a l_b + s\overline{m}^b\nabla_a m_b
\end{aligned}
$$

where r is the boost-weight given in (3.185) and s is the spin-weight given in (3.184). Contracting with appropriate null vectors l^a, n^a, m^a and $\overline{m}^a$ will yield the expression in (3.194).

The types of weighted differential operators can be obtained directly from (3.195). Thus,

$$
\begin{aligned}
þ &: \{1, 1\}\text{-operator} \\
þ' &: \{-1, -1\}\text{-operator} \\
ð &: \{1, -1\}\text{-operator}
\end{aligned}
$$

$$\eth' : \{-1, 1\}\text{-operator.} \tag{3.198}$$

Because this formalism is dependent on weighted quantities alone, the spin coefficients β, β', ϵ, and ϵ' in (3.186) and the intrinsic derivatives D, D', δ, δ' in (3.187) participate in the formalism only via the relations (3.194). Consequently, one deals with a greatly reduced number of spin coefficients compared with the NP formalism. Indeed, the GHP formalism consists entirely of the first eight weighted spin coefficients in (3.186) and the four weighted differential operators in (3.194), together with the operations of complex conjugation and 'priming'. A double application of the prime operation on a $\{p, q\}$-scalar effectively leaves it invariant:

$$(\eta')' = (-1)^{(p+q)}\eta. \tag{3.199}$$

This is because $(p+q)$ will always be even here.

Let us define the following conjugated weighted differential operators differential operators by

$$\begin{aligned} \bar{þ} &= þ \\ \bar{þ}' &= þ' \\ \bar{\eth} &= \eth \\ \bar{\eth}' &= \eth'. \end{aligned} \tag{3.200}$$

As the complex conjugate of a quantity of type $\{p, q\}$ is a quantity of type $\{\bar{q}, \bar{p}\}$, we have

$$\begin{aligned} \overline{þ\eta} &= \bar{þ}\bar{\eta} \\ \overline{\eth\eta} &= \bar{\eth}\bar{\eta}. \end{aligned} \tag{3.201}$$

The effect of applying the prime operation to a quantity of type $\{p, q\}$ is to replace it with a quantity of type $\{-p, -q\}$. Also,

$$\begin{aligned} (þ\eta)' &= þ'\eta' \\ (þ'\eta)' &= þ\eta' \\ (\eth\eta)' &= \eth'\eta' \\ (\eth'\eta)' &= \eth\eta'. \end{aligned} \tag{3.202}$$

3.8.4 *The Geroch–Held–Penrose field equations*

It is a straightforward exercise to obtain the GHP form of the field equations given originally in (3.132). Then from (3.132) (1), (7), (11), (14), (6) and

(18) we have, respectively,

$$
\begin{aligned}
\text{þ}\rho - \eth'\kappa &= \rho^2 + \sigma\bar{\sigma} - \bar{\kappa}\tau - \tau'\kappa + \Phi_{00} \\
\text{þ}\sigma - \eth\kappa &= (\rho + \bar{\rho})\sigma - (\tau + \bar{\tau}')\kappa + \Psi_0 \\
\text{þ}\tau - \text{þ}'\kappa &= (\tau - \bar{\tau}')\rho + (\bar{\tau} - \tau')\sigma + \Psi_1 + \Phi_{01} \\
\eth\rho - \eth\sigma &= (\rho - \bar{\rho})\tau + (\bar{\rho}' - \rho')\kappa - \Psi_1 + \Phi_{01} \\
\eth\tau - \text{þ}'\sigma &= -\rho'\sigma - \bar{\sigma}\rho + \tau^2 + \kappa\bar{\kappa}' + \Phi_{02} \\
\text{þ}'\rho - \eth'\tau &= \rho\bar{\rho}' + \sigma\sigma' - \tau\bar{\tau} - \kappa\kappa' - \Psi_2 - 2\Lambda. \qquad (3.203)
\end{aligned}
$$

There exist another six equations which correspond to (3.132)(4), (10), (13), (12), (3) and (17). These are obtained by applying the prime operation to (3.203) and taking into consideration (3.164) and (3.165). They are given, respectively, by

$$
\begin{aligned}
\text{þ}'\rho' - \eth\kappa' &= \rho'^2 + \sigma'\bar{\sigma}' - \bar{\kappa}'\tau' - \tau\kappa' + \Phi_{22} \\
\text{þ}'\sigma' - \eth'\kappa' &= (\rho' + \bar{\rho}')\sigma' - (\tau' + \bar{\tau})\kappa' + \Psi_4 \\
\text{þ}'\tau' - \text{þ}\kappa' &= (\tau' - \bar{\tau})\rho' + (\bar{\tau}' - \tau)\sigma' + \Psi_3 + \Phi_{21} \\
\eth'\rho' - \eth\sigma' &= (\rho' - \bar{\rho}')\tau' + (\bar{\rho}' - \rho)\kappa' - \Phi_3 + \Phi_{21} \\
\eth'\tau' - \text{þ}\sigma' &= -\rho\sigma' - \bar{\sigma}\rho' + \tau'^2 + \kappa'\bar{\kappa} + \Phi_{20} \\
\text{þ}\rho' - \eth\tau' &= \rho'\bar{\rho} + \sigma'\sigma - \tau'\bar{\tau}' - \kappa'\kappa - \Psi_2 - 2\Lambda. \qquad (3.204)
\end{aligned}
$$

To illustrate the method whereby we transform from the NP to the GHP field equations consider, for example, (3.132)(1). By employing (3.163) and (3.187) this can be written as

$$D\rho - \delta'\kappa = (\rho^2 + \sigma\bar{\sigma}) + (\epsilon + \bar{\epsilon})\rho - \bar{\kappa}\tau - \kappa(-3\beta' + \bar{\beta} + \tau') + \Phi_{00}$$

or, after some slight rearranging,

$$(D - \epsilon - \bar{\epsilon})\rho - (\delta' + 3\beta' - \bar{\beta})\kappa = \rho^2 + \sigma\bar{\sigma} - \bar{\kappa}\tau - \tau'\kappa + \Phi_{00}.$$

Recalling that ρ is a $\{1, 1\}$-scalar and κ is a $\{3, 1\}$-scalar (see 3.190) then with the aid of (3.194) the result follows.

It is apparent if one glances at (3.203) and (3.204) that six field equations are missing; namely, (3.132)(2), (5), (8), (9), (15) and (16). The reason for their absence is that they are unweighted quantities and do not participate explicitly in the formalism. However, these equations do contribute towards forming the commutation relations for the differential operators þ, þ$'$, $\eth$ and $\eth'$. The six unweighted field equations are obtained by combining the following three equations with their primed versions:

$$\begin{aligned}
þ\beta' + ð'\epsilon &= [\rho - \epsilon - (q-1)\overline{\epsilon}]\beta' + [\tau' - \beta' - (q-1)\overline{\beta}]\epsilon \\
&\quad -\beta\overline{\sigma} + \rho\tau' - \kappa\sigma' - \overline{\kappa}\epsilon' - \Phi_{10} \\
þ\beta - ð\epsilon &= [\overline{\rho} - (q+1)\overline{\epsilon}]\beta + [-\overline{\tau}' - (q-1)\overline{\beta}']\epsilon \\
&\quad -(\beta' + \tau')\sigma + (\rho' + \epsilon')\kappa + \Psi_1 \\
þ\epsilon' + þ'\epsilon &= -[\epsilon + (q+1)\overline{\epsilon}]\epsilon' - [\epsilon' - (q-1)\overline{\epsilon}']\epsilon \\
&\quad +(\tau - \overline{\tau}')\beta' + (\tau' - \overline{\tau})\beta + \tau\tau' - \kappa\kappa' - \Psi_2 - \Phi_{11} + \Lambda.
\end{aligned} \tag{3.205}$$

The commutation relations when applied to a $\{p,q\}$-scalar η are given by the three equations

$$\begin{aligned}
(þþ' - þ'þ)\eta &= [(\overline{\tau} - \tau')ð + (\tau - \overline{\tau}')ð' \\
&\quad -p(\kappa\kappa' - \tau\tau' + \Psi_2 + \Phi_{11} - \Lambda) \\
&\quad -q(\overline{\kappa\kappa}' - \overline{\tau\tau}' + \overline{\Psi}_2 + \Phi_{11} - \Lambda)]\eta
\end{aligned} \tag{3.206}$$

$$\begin{aligned}
(þð - ðþ)\eta &= [\overline{\rho}ð + \sigma ð' - \overline{\tau}'þ - \kappa þ' \\
&\quad -p(\rho'\kappa - \tau'\sigma + \Psi_1) \\
&\quad -q(\overline{\sigma}'\overline{\kappa} - \overline{\rho\tau}' + \Phi_{01})]\eta
\end{aligned} \tag{3.207}$$

$$\begin{aligned}
(ðð' - ð'ð)\eta &= [(\overline{\rho}' - \rho')þ + (\rho - \overline{\rho})þ' \\
&\quad +p(\rho\rho' - \sigma\sigma' + \Psi_2 - \Phi_{11} - \Lambda) \\
&\quad -q(\overline{\rho\rho}' - \overline{\sigma\sigma}' + \overline{\Psi}_2 - \Phi_{11} - \Lambda)]\eta
\end{aligned} \tag{3.208}$$

in addition to another three equations obtained by applying the prime operation, complex conjugation and the combination of both to (3.207). Great care must be taken when applying these procedures. To some extent, we have already encountered the difference in type resulting from the quantities $\overline{\eta}$, η' and $\overline{\eta}'$. They are explicitly categorised in terms of their weights as follows:

$$\begin{aligned}
\eta &\mapsto \overline{\eta} \Rightarrow \{p,q\} \mapsto \{q,p\} \\
\eta &\mapsto \eta' \Rightarrow \{p,q\} \mapsto \{-p,-q\} \\
\eta &\mapsto \overline{\eta}' \Rightarrow \{p,q\} \mapsto \{-q,-p\}.
\end{aligned} \tag{3.209}$$

To demonstrate the significance of (3.209), we will derive (3.206).

From the commutation relations (3.128) consider relation (1). Replacing the intrinsic derivatives with the weighted operators defined by (3.194), and recalling that $\gamma = -\epsilon'$, $\pi = -\tau'$ and $\alpha = -\beta'$, we have

$$\begin{aligned}
þþ' - þ'þ &= (\overline{\tau} - \tau')ð + (\tau - \overline{\tau}')ð' \\
&-[(p-1)\epsilon' + (q-1)\overline{\epsilon}']þ - [(p+1)\epsilon + (q+1)\overline{\epsilon}]þ \\
&-p\{-þ\epsilon' - þ'\epsilon - (\epsilon' + \overline{\epsilon}')\epsilon - (\epsilon + \overline{\epsilon})\epsilon' - (\overline{\tau} - \tau')\beta + (\tau - \overline{\tau}')\beta'\} \\
&-q\{-þ\overline{\epsilon}' - þ'\overline{\epsilon} - (\epsilon' + \overline{\epsilon}')\overline{\epsilon} - (\epsilon + \overline{\epsilon})\overline{\epsilon}' - (\tau - \overline{\tau}')\overline{\beta} + (\overline{\tau} - \tau')\overline{\beta}'\}.
\end{aligned}$$

Substituting (3.205) and its complex conjugate into the above equation, and also noting that the weights of þ and þ′ are types $\{1,1\}$ and $\{-1,-1\}$ respectively, yields the result.

The Bianchi identities (3.139) exhibit a rather simple and economical form in the GHP formalism. They are readily obtainable with the use of (3.194), (3.198), (3.192) and (3.193):

$$\begin{aligned}
þ\Psi_1 - ð'\Psi_0 - þ\Phi_{01} + ð\Phi_{00} &= -\tau'\Psi_0 + 4\rho\Psi_1 - 3\kappa\Psi_2 \\
&+\overline{\tau}'\Phi_{00} - 2\overline{\rho}\Phi_{01} - 2\sigma\Phi_{10} + 2\kappa\Phi_{11} + \overline{\kappa}\Phi_{02},
\end{aligned} \tag{3.210}$$

$$\begin{aligned}
þ\Psi_2 - ð\Psi_1 - ð'\Phi_{01} + þ'\Phi_{00} + 2þ\Lambda &= \sigma'\Psi_0 - 2\tau'\Psi_1 + 3\rho\Psi_2 - 2\kappa\Psi_3 \\
&+\overline{\rho}'\Phi_{00} - 2\overline{\tau}\Phi_{01} - 2\tau\Phi_{10} + 2\rho\Phi_{11} + \overline{\sigma}\Phi_{02},
\end{aligned} \tag{3.211}$$

$$\begin{aligned}
þ\Psi_3 - ð'\Psi_2 - þ\Phi_{21} + ð\Phi_{20} - 2ð\Lambda &= 2\sigma'\Psi_1 - 3\tau'\Psi_2 + 2\rho\Psi_3 - \kappa\Psi_4 \\
&- 2\rho'\Phi_{10} + 2\tau'\Phi_{11} + \overline{\tau}'\Phi_{20} - 2\overline{\rho}\Phi_{21} + \overline{\kappa}\Phi_{22},
\end{aligned} \tag{3.212}$$

$$\begin{aligned}
þ\Psi_4 - ð'\Psi_3 - ð'\Phi_{21} + þ'\Phi_{20} &= +3\sigma'\Psi_2 - 4\tau'\Psi_3 + \rho\Psi_4 \\
&- 2\kappa'\Phi_{10} + 2\sigma'\Phi_{11} + \overline{\rho}'\Phi_{20} - 2\overline{\tau}\Phi_{21} + \overline{\sigma}\Phi_{22},
\end{aligned} \tag{3.213}$$

$$\begin{aligned}
þ\Phi_{11} + þ'\Phi_{00} - ð\Phi_{10} - ð'\Phi_{01} + 3þ\Lambda &= (\rho' + \overline{\rho}')\Phi_{00} + 2(\rho + \overline{\rho})\Phi_{11} \\
&- (\tau' + 2\overline{\tau})\Phi_{01} - (2\tau + \overline{\tau}')\Phi_{10} \\
&- \overline{\kappa}\Phi_{12} - \kappa\Phi_{21} + \sigma\Phi_{20} + \overline{\sigma}\Phi_{02},
\end{aligned} \tag{3.214}$$

$$\begin{aligned}
þ\Phi_{12} + þ'\Phi_{01} - ð\Phi_{11} - ð'\Phi_{02} + 3ð\Lambda &= (\rho' + 2\overline{\rho}')\Phi_{01} + (2\rho + \overline{\rho})\Phi_{12} \\
&- (\tau' + \overline{\tau})\Phi_{02} - 2(\tau + \overline{\tau}')\Phi_{11} \\
&- \overline{\kappa}'\Phi_{00} - \kappa\Phi_{22} + \sigma\Phi_{21} + \overline{\sigma}'\Phi_{10},
\end{aligned} \tag{3.215}$$

together with their primed versions. Equations (3.214) and (3.215) form the contracted Bianchi identities (see (3.97)).

It is worth mentioning a little regarding the asterisk operation of Sec. 3.7.3. The way in which the null tetrad and spin coefficients etc. transform under the asterisk operation has already been detailed in that section, and so it remains only to discuss it with respect to the GHP formalism.

For a weighted quantity such as the $\{p, q\}$-scalar η, we have

$$(\eta^*)^* = (-1)^q \eta \tag{3.216}$$

$$(\eta')^* = (-1)^q (\eta^*)' \tag{3.217}$$

$$\overline{(\eta^*)} = (-i)^{p+q} (\overline{\eta}')^* \tag{3.218}$$

where η^* is a $\{p, -q\}$-scalar. By employing (3.194), (3.186), (3.168), (3.171) and (3.172), we find that

$$\begin{aligned} þ^* &= ð \\ þ'^* &= -ð' \\ ð^* &= -þ \\ ð'^* &= þ'. \end{aligned} \tag{3.219}$$

The individual 'sets' of equations which comprise the GHP field equations, commutation relations, and Bianchi identities remain unaltered in form under the asterisk operation.

3.8.5 *A brief note on the modified GHP formalism*

A greater simplification of the GHP formalism can be achieved (Held, 1974, 1975) when the differential operators ð and þ are modified. That is, by introducing the operators $\tilde{ð}$, $\tilde{ð}'$, and $\tilde{þ}'$ acting on a $\{p, q\}$ quantity, η, (3.194) becomes (see Stewart and Walker, 1974),

$$\begin{aligned} \tilde{ð}\eta &= \left(\overline{\rho}^{-1} ð + \frac{q\tau}{\rho}\right)\eta \\ \tilde{ð}'\eta &= \left(\rho^{-1} ð' + \frac{p\overline{\tau}}{\overline{\rho}}\right)\eta \\ \tilde{þ}\eta &= \left[þ' - \overline{\tau}\tilde{ð} - \tau\tilde{ð}' + \tau\overline{\tau}\left(\frac{p}{\overline{\rho}} + \frac{q}{\rho}\right) + \frac{1}{2}\left(\frac{p\Psi_2}{\rho} + \frac{q\overline{\Psi}_2}{\overline{\rho}}\right)\right]\eta. \end{aligned}$$

The modified operators possess the property that their commutators with þ are 'proportional' to þ, i.e. if $þ\eta^\circ = 0$ we have

$$[þ, \tilde{þ}']\eta^\circ = 0$$

$$[þ, \tilde{ð}]\eta^\circ = 0$$

$$[þ, \tilde{ð}']\eta^\circ = 0,$$

where the degree sign is indicative of the fact that the function is annihilated by þ. Unfortunately, modifying the operators in this manner no longer allows us to adopt the prime operation (Sec. 3.7.2) nor the asterisk operation (Sec. 3.7.3). Although this formalism works very well in the case of algebraically special space-times, cannot be applied to metrics which are algebraically general.

3.9 Goldberg–Sachs theorem

This theorem (Goldberg and Sachs, 1962) states that the Riemann tensor in empty space-time ($R_{ab} = 0$) is algebraically special *if and only if* a principal null direction defines a congruence of null geodesics which are shear-free; that is,

$$\kappa = \sigma = 0 \Longleftrightarrow \Psi_0 = \Psi_1 = 0. \tag{3.220}$$

(Note: if $\kappa = 0$ the null vector l^a will be tangent to a congruence of null geodesics, and the quantity σ is called the *shear*.)

Let us first prove the right-left implication. By putting $\Psi_0 = \Psi_1 = 0$ into the Bianchi identities (3.137) they reduce to

$$\begin{aligned}
3\sigma\Psi_2 &= 0\\
\delta\Psi_2 - 3\tau\Psi_2 + 2\sigma\Psi_3 &= 0\\
\Delta\Psi_2 - \delta\Psi_3 + 3\mu\Psi_2 + 2(\tau - \beta)\Psi_3 + \sigma\Psi_4 &= 0\\
3\kappa\Psi_2 &= 0\\
D\Psi_2 + 2\kappa\Psi_3 - 3\rho\Psi_2 &= 0\\
D\Psi_3 - \bar{\delta}\Psi_2 + \kappa\Psi_4 + 2(\epsilon - \rho)\Psi_3 - 3\pi\Psi_2 &= 0.
\end{aligned}$$

Clearly if $\sigma \neq 0$ or $\kappa \neq 0$ then $\Psi_2 = \Psi_3 = \Psi_4 = 0$; that is, the space is flat.

Let us next prove the left-right implication. It is possible to put $\epsilon = 0$ by combining the spin transformation (3.150) with a scaling $l^a \mapsto al^a$ for a suitably chosen real function a. This together with the assumption $\kappa = \sigma = 0$ reduces the relevant NP field equations (3.132) to

$$D\rho - \rho^2 = 0 \tag{3.221}$$

$$\Psi_0 = 0 \tag{3.222}$$

$$D\tau - (\tau + \overline{\pi})\rho - \Psi_1 = 0 \tag{3.223}$$

$$D\beta - \beta\overline{\rho} - \Psi_1 = 0 \tag{3.224}$$

$$\delta\rho - (\overline{\alpha} + \beta)\rho - (\rho - \overline{\rho})\tau + \Psi_1 = 0. \tag{3.225}$$

Because of (3.222) we need only consider the first two Bianchi equations in (3.137) which reduce to

$$D\Psi_1 - 4\rho\Psi_1 = 0 \tag{3.226}$$

$$\delta\Psi_1 - 2(2\tau + \beta)\Psi_1 = 0, \tag{3.227}$$

and the second commutation equation in (3.128)

$$(\delta D - D\delta)\phi = [(\overline{\alpha} + \beta - \overline{\pi})D - (\overline{\rho} + \epsilon - \overline{\epsilon})\delta]\phi. \tag{3.228}$$

The choice of this particular commutation equation is implied by (3.226) and (3.227).

Notice that freedom still remains in choosing the null vector n^a. This is removed if we employ (3.139):

$$\begin{aligned} l^a &\mapsto l^a \\ n^a &\mapsto n^a + \overline{c}m^a + c\overline{m}^a + c\overline{c}l^a \\ m^a &\mapsto m^a + cl^a \\ \overline{m}^a &\mapsto \overline{m}^a + \overline{c}l^a. \end{aligned}$$

Clearly l^a remains invariant under this rotation, and maintains the relation $\epsilon = 0$. Provided $\rho \neq 0$ the complex parameter c can be chosen such that $\tau = 0$. Hence, (3.223) becomes

$$\Psi_1 = -\overline{\pi}\rho. \tag{3.229}$$

Also, (3.226) and (3.227) are rewritten as

$$\begin{aligned} D \ln \Psi_1 &= 4\rho \\ \delta \ln \Psi_1 &= 2\beta, \end{aligned}$$

or, on differentiating,

$$\delta D \ln \Psi_1 = 4\delta\rho \tag{3.230}$$

$$D\delta \ln \Psi_1 = 2D\beta. \tag{3.231}$$

Subtracting (3.231) from (3.230) and comparing the result with (3.224) and (3.225) yields

$$(\delta D - D\delta) \ln \Psi_1 = 4(\overline{\alpha} + \beta)\rho - 2\beta\overline{\rho} - 6\Psi_1. \tag{3.232}$$

Letting $\phi = \ln \Psi_1$ in (3.228) gives

$$(\delta D - D\delta) \ln \Psi_1 = 4(\overline{\alpha} + \beta)\rho - 2\beta\overline{\rho} - 4\rho\overline{\pi}, \tag{3.233}$$

which on subtracting from (3.232) results in

$$\Psi_1 = \frac{2}{3}\rho\overline{\pi}.$$

By comparing this with (3.229) we see that $\Psi_1 = \overline{\pi} = 0$; and as it was assumed that $\rho \neq 0$, the proof is complete.

Notice, however, that if we had allowed ρ to vanish in the second part of the proof then $\Psi_1 = 0$ follows straight away from (3.225) and again this completes the proof.

3.10 Exercises

3.1 Obtain equation (3.60).

3.2 Show that the last column in Table 2.2 is equivalent to the last column in Table 2.3.

3.3 Obtain equations (3.79) and (3.80).

3.4 Obtain equations (3.84)–(3.88).

3.5 Show, using equation (3.97), that in vacuum

$$\nabla^{AA'}\Psi_{ABCD} = 0.$$

Also, that in the non-vacuum case

$$\nabla^{A}_{B'}\Psi_{ABCD} = 4\pi\gamma\nabla^{A'}_{(B}T_{CD)A'B'},$$

where $T_{ABA'B'}$ is the energy-momentum spinor (see Sec. 3.2.2). What can be inferred about the derivative of the energy-momentum tensor?

3.6 Obtain the dyad form of the spinor $\nabla_{AA'}(\zeta_C \chi^C{}_B)$.

3.7 Referring to equations (3.112) and (3.113), repeat the calculation which led to (3.111) for a non-normalised basis, i.e. show that

$$\gamma_{\mathbf{AA'BC}} - \gamma_{\mathbf{AA'CB}} = \epsilon_{\mathbf{BC}} \Omega^{-1} \nabla_{\mathbf{AA'}} \Omega.$$

3.8 Verify (3.118) and (3.120).

3.9 Establish (3.128) by transvecting (3.121) with appropriate null tetrad vectors.

3.10 Derive the Bianchi identities (3.137).

3.11 Derive a selection of the transformation equations in Sec. 3.6.

3.12 Obtain equations (3.188) and (3.189).

3.13 Obtain equations(3.203) and (3.204).

3.14 Show that for empty space-time ($R_{ab} = 0$) the condition $\Psi_3 = \Psi_4 = 0$ is necessary and sufficient for $\nu = \lambda = 0$.

Chapter 4

Lanczos spinor

4.1 Introduction

During the early part of the 1960's Cornelius Lanczos (1893-1974), while analysing the self-dual part of the Riemann tensor R_{abcd} in four dimensions, discovered an essentially new tensor H_{abc} of third-order. Although not mentioned explicitly in his 1962 paper, the Weyl tensor could in fact be generated from this new tensor, making it, therefore, a much more fundamental geometrical object. Lanczos' primary motivation in concentrating on the self-dual part of the Riemann tensor is simple to understand: not to investigate it deprives us of important knowledge and a thorough understanding of Riemannian geometry. Another quite unconnected reason was due to Einstein's neglect of it in preference to analysing the anti-self-dual part (Einstein, 1926). However, Einstein quickly abandoned his analysis of the anti-self-dual for reasons which will become apparent.

Let us define the anti-self-dual and self-dual parts of the Riemann tensor respectively by

$$A_{abcd} = R_{abcd} -^{*} R^{*}{}_{abcd} \tag{4.1}$$

and

$$S_{abcd} = R_{abcd} +^{*} R^{*}{}_{abcd}. \tag{4.2}$$

We can then write the Riemann tensor as

$$R_{abcd} = \frac{1}{2}(A_{abcd} + S_{abcd}). \tag{4.3}$$

As the Riemann tensor has 20 independent components, the right-hand side of (4.3) must have 20 independent components. The number of components, however, is not split equally between the anti-self-dual and self-dual parts. The anti-self-dual has 9 components whilst the self-dual has 11. The type of splitting in (4.1) and (4.2) can be attributed to Rainich (Rainich, 1925). Using this method Einstein obtained the anti-self-dual tensor A_{abcd} in terms of the Ricci tensor, Ricci scalar and metric tensor:

$$A_{abcd} = (R_{ac} - \frac{1}{4}R\,g_{ac})g_{bd} + (R_{bd} - \frac{1}{4}R\,g_{bd})g_{ac}$$
$$-(R_{ad} - \frac{1}{4}R\,g_{ad})g_{bc} - (R_{bc} - \frac{1}{4}R\,g_{bc})g_{ad}. \tag{4.4}$$

From this expression Einstein was able to deduce the field equations

$$R_{ab} - \frac{1}{4}R\,g_{ab} = -KT_{ab} \tag{4.5}$$

where K is a constant and T_{ab} is the energy-momentum tensor. Einstein's hope was to model the structure of a stable electron. However, if one were to contract (4.5) with g^{ab} one would find that the equation would contain 9 independent components, whereas the trace of the energy-momentum tensor should have 10 independent components leading to an ambiguity of the charge for a given field arbitrarily distributed through space-time. This consequently forced Einstein to terminate any further analysis of the anti-self-dual part of the Riemann tensor.

4.2 Lanczos' Lagrangian

In his analysis of the self-dual tensor (4.2) Lanczos decided to approach the problem via a variational principle. The first task was to construct a Lagrangian which contained only first-order derivatives. The Lagrange multipliers would then have some physical importance and the field equations would be easily soluble.

One of the most important aspects of Lanczos' Lagrangian was that the dual ${}^*R^*{}_{abcd}$ and the metric tensor g_{ab} were treated as independent quantities. He was justified in doing this by introducing three important subsidiary equations:

$${}^*R^{*abcd}{}_{;d} = 0, \tag{4.6}$$

$$\Gamma^b_{ac} - \left\{ {b \atop ac} \right\} = 0, \tag{4.7}$$

$$R_{ab} + \Gamma^c_{bc,a} - \Gamma^c_{ab,c} + \Gamma^c_{ad}\Gamma^d_{bc} - \Gamma^c_{ab}\Gamma^d_{dc} = 0. \tag{4.8}$$

These are respectively: the Bianchi identities; the relationship between the Christoffel symbols and affine connexion; and, lastly, the contracted

Riemann tensor. If we know the variations of ${}^*R^{*abcd}$ then we know the variations of g_{ab}. However, the full Riemann tensor is not required and it is sufficient to consider the Ricci tensor alone, which establishes the metric tensor. Also, on account of the presence of (4.7), equation (4.8) is of first-order. Thus following from above Lanczos introduces his Lagrangian

$$\begin{aligned} L = \frac{1}{8}\,{}^*R^{*abcd}R_{abcd} + H_{abc}\,{}^*R^{*abcd}{}_{;d} + P^{ab}{}_c(\Gamma^c_{ab} - \{{}^{\,c}_{ab}\}) \\ +\rho^{ab}(R_{ab} + \Gamma^c_{bc,a} - \Gamma^c_{ab,c} + \Gamma^c_{ad}\,\Gamma^d_{bc} - \Gamma^c_{ab}\,\Gamma^d_{dc}) \end{aligned} \qquad (4.9)$$

where the canonical variables g_{ab}, $\Gamma^c{}_{ab}$ and ${}^*R^{*abcd}$ have the conjugates $P^{ab}{}_c$, ρ^{ab} and H_{abc}. The Lagrange multiplier H_{abc} is now commonly referred to as the Lanczos tensor. In his original paper, Lanczos called this object the *spintensor* which was presumably so named because of the connection between Dirac's equation for the electron and the system of equations obtained by 'splitting' the Riemann tensor. Because H_{abc} is skew-symmetric in a, b it has 24 algebraically independent components; whereas ${}^*R^{*abcd}$ has 20 independent components. In view of this, to restrict the number of independent components of H_{abc} to 20, Lanczos introduced, without any loss of generality, the condition

$${}^*H^{ad}{}_a = \frac{1}{2}H_{abc}\,\epsilon^{abcd} = 0, \qquad (4.10)$$

which is equivalent to the cyclic condition

$$H_{abc} + H_{bca} + H_{cab} = 0. \qquad (4.11)$$

Both the Lagrange multipliers $P^{ab}{}_c$ and ρ^{ab} are symmetric in a,b.

Now we wish to vary (4.9) with respect to the canonical variable ${}^*R^{*abcd}$. It will be useful to recall that the symmetry properties of ${}^*R^{*abcd}$ are the same as R^{abcd},

$${}^*R^{*abcd} = -{}^*R^{*bacd} = -{}^*R^{*abdc} = {}^*R^{*cdab}. \qquad (4.12)$$

The cyclic condition also holds,

$${}^*R^{*abcd} + {}^*R^{*bcad} + {}^*R^{*cabd} = 0. \qquad (4.13)$$

It will also be of use to decompose ρ^{ab} into its trace and trace-free parts

$$\rho^{ab} = Q^{ab} + qg^{ab}, \qquad (4.14)$$

where q is a scalar quantity defined by

$$q = \frac{1}{4}\rho^{ab}g_{ab} \qquad (4.15)$$

and, of course,

$$Q^{ab}g_{ab} = Q^a{}_a = 0. \tag{4.16}$$

Before we can vary (4.9) it is necessary to make some slight modification to assist the variation. By writing the second term as

$$H_{abc}\,{}^*R^{*abcd}{}_{;d} = (H_{abc}\,{}^*R^{*abcd})_{;d} - H_{abc;d}\,{}^*R^{*abcd} \tag{4.17}$$

we assume that the first term on the right-hand side vanishes by construction. Because $\Gamma^a{}_{bc}$ and $P^{ab}{}_c$ vanish when varied with respect to ${}^*R^{*abcd}$ we can rewrite (4.9), with the aid of (4.14) and (4.17), as

$$\begin{aligned} L' = {} & \frac{1}{8}\,{}^*R^{*abcd}R_{abcd} - H_{abc;d}\,{}^*R^{*abcd} \\ & +(Q^{ab} + qg^{ab})\,R_{ab}. \end{aligned} \tag{4.18}$$

Now, in terms of the dual ${}^*R^*{}_{abcd}$ the Einstein tensor can be written as

$$R_{ac} - \frac{1}{2}R\,g_{ac} = {}^*R^*{}_{abc}{}^b = {}^*R^{*ebfd}g_{ea}g_{fc}g_{bd}. \tag{4.19}$$

Then the last term in (4.18) is given by

$$\begin{aligned} (Q^{ac} + qg^{ac})R_{ac} &= {}^*R^{*ebfd}(Q^{ac} + qg^{ac})g_{ea}g_{fc}g_{bd} \\ &\quad +\frac{1}{2}R\,g_{ac}(Q^{ac} + qg^{ac}) \\ &= {}^*R^{*abcd}(Q_{ac} + qg_{ac})g_{bd} + 2qR. \end{aligned}$$

Contracting (4.19) with g^{ac} yields

$$R = -{}^*R^{*abcd}g_{ac}g_{bd},$$

giving finally,

$$(Q^{ac} + qg^{ac})R_{ac} = {}^*R^{*abcd}(Q_{ac} - qg_{ac})g_{bd}. \tag{4.20}$$

Substituting (4.20) into (4.18) and varying the Lagrangian with respect to ${}^*R^{*abcd}$ yields

$$\frac{1}{4}R_{abcd} - H_{abc;\,d} + Q_{ac}\,g_{bd} - qg_{ac}g_{bd} = 0. \tag{4.21}$$

However, we cannot leave this expression as it stands. We must now consider the algebraic symmetry properties of R^{abcd}. Thus after employing these symmetry properties the third term in (4.21) can be written

$$\begin{aligned}Q_{ac}g_{bd} &= \frac{1}{2}[Q_{ac}g_{bd} + Q_{ca}g_{db}]\\ &= \frac{1}{4}[Q_{ac}g_{bd} - Q_{bc}g_{ad} + Q_{ca}g_{db} - Q_{bd}g_{ac}]\\ &= \frac{1}{8}[Q_{ac}g_{bd} - Q_{ad}g_{bc} - Q_{bc}g_{ad} + Q_{bd}g_{ac}\\ &\quad + Q_{ca}g_{db} - Q_{da}g_{db} - Q_{cb}g_{da} + Q_{db}g_{ca}].\end{aligned}$$

After taking into account the symmetry properties of Q_{ab} we get

$$Q_{ac}g_{bd} = \frac{1}{4}(Q_{ac}g_{bd} - Q_{ad}g_{bc} - Q_{bc}g_{ad} + Q_{bd}g_{ac}). \tag{4.22}$$

The fourth term in (4.20) can be written in a similar fashion after taking into account the symmetry properties of g_{ab},

$$qg_{ac}g_{bd} = \frac{1}{2}q(g_{ac}g_{bd} - g_{ad}g_{bc}). \tag{4.23}$$

Because $H_{abc;d}$ possess the same skew-symmetry properties on a, b as well as the same cyclic permutation as the Riemann tensor we have

$$\begin{aligned}H_{abc;d} &= \frac{1}{2}(H_{abc;d} + H_{cda;b})\\ &= \frac{1}{4}(H_{abc;d} - H_{abd;c} + H_{cda;b} - H_{cdb;a}).\end{aligned} \tag{4.24}$$

Substituting (4.22), (4.23) and (4.24) into (4.21) we get

$$\begin{aligned}R_{abcd} &= H_{abc;d} - H_{abd;c} + H_{cda;b} - H_{cdb;a}\\ &\quad -(Q_{ac} - qg_{ac})g_{bd} - (Q_{bd} - qg_{bd})g_{ac}\\ &\quad +(Q_{ad} - qg_{ad})g_{bc} + (Q_{bc} - qg_{bc})g_{ad}.\end{aligned} \tag{4.25}$$

By making certain contractions with the metric tensor one obtains direct relationships between Q_{ab}, the scalar q and other known quantities. Thus contracting (4.25) with g^{ad} yields the *Ricci–Lanczos equations*:

$$R_{bc} = H^{a}{}_{bc;a} - H^{a}{}_{ba;c} + H_{c}{}^{a}{}_{a;b} - H_{c}{}^{\alpha}{}_{b;a} - 2Q_{bc} + 6qg_{bc}. \tag{4.26}$$

Contracting this with g^{bc} gives

$$\begin{aligned}R &= H^{ab}{}_{b;a} - H^{ab}{}_{a;b} + H^{ba}{}_{a;b} - H^{ba}{}_{b;a} + 24q\\ &= 4H^{ab}{}_{b;a} + 24q.\end{aligned} \tag{4.27}$$

After some slight rearranging of (4.26) and (4.27) we can write

$$Q_{bc} - q g_{bc} = H^{a}{}_{(bc);a} + H_{(c|a|}{}^{a}{}_{;b)} - \frac{1}{3} H^{ad}{}_{d;a} g_{bc} - \frac{1}{2} R_{bc} + \frac{1}{12} R\, g_{bc}.$$

Substituting this into (4.25) yields the *Riemann–Lanczos equations*:

$$\begin{aligned} R_{abcd} = {} & H_{abc;d} + H_{cda;b} + H_{bad;c} + H_{dcb;a} \\ & + [H^{e}{}_{(bd);e} + H_{(b|e|}{}^{e}{}_{;d)}] g_{ac} \\ & - [H^{e}{}_{(ad);e} + H_{(a|e|}{}^{e}{}_{;d)}] g_{bc} \\ & - [H^{e}{}_{(bc);e} + H_{(b|e|}{}^{e}{}_{;c)}] g_{ad} \\ & - \frac{2}{3} H^{ef}{}_{f;e} [g_{ac} g_{bd} - g_{ad} g_{bc}] \\ & - \frac{1}{2} (R_{ac} g_{bd} + R_{bd} g_{ac} - R_{ad} g_{bc} - R_{bc} g_{ad}) \\ & + \frac{1}{6} (g_{ac} g_{bd} - g_{ad} g_{bc}) R. \end{aligned} \tag{4.28}$$

If we now compare equation (4.28) with the well known equation which splits the Riemann tensor into the Weyl tensor and parts which involve the Ricci tensor and corresponding scalar,

$$\begin{aligned} R_{abcd} = {} & C_{abcd} - \frac{1}{2} (R_{ac} g_{bd} + R_{bd} g_{ac} - R_{ad} g_{bc} - R_{bc} g_{ad}) \\ & + \frac{1}{6} R\, (g_{ac} g_{bd} - g_{ad} g_{bc}), \end{aligned} \tag{4.29}$$

we obtain a direct relationship between the Weyl tensor and the Lanczos tensor, namely

$$\begin{aligned} C_{abcd} = {} & H_{abc;d} + H_{cda;b} + H_{bad;c} + H_{dcb;a} \\ & + [H^{e}{}_{(ac);e} + H_{(a|e|}{}^{e}{}_{;c)}] g_{bd} \\ & + [H^{e}{}_{(bd);e} + H_{(b|e|}{}^{e}{}_{;d)}] g_{ac} \\ & - [H^{e}{}_{(ad);e} + H_{(a|e|}{}^{e}{}_{;d)}] g_{bc} \\ & - [H^{e}{}_{(bc);e} + H_{(b|e|}{}^{e}{}_{;c)}] g_{ad} \\ & - \frac{2}{3} H^{ef}{}_{f;e} [g_{ac} g_{bd} - g_{ad} g_{bc}]. \end{aligned} \tag{4.30}$$

These are the Weyl–Lanczos equations. They imply that the Weyl tensor can no longer be thought of as the fundamental object it once was, but merely as a tensor generated by a new more fundamental tensor: the Lanczos tensor. Lanczos himself placed the importance of this new tensor between the metric tensor and Riemann tensor. This is of course justified. Yet it initially appears rather puzzling then why there is little material in the relevant literature on the Lanczos tensor compared with the vast quantity encountered on the Weyl tensor. The answer to this question will become apparent in the following sections.

4.3 Lanczos' gauge conditions

The Weyl tensor has 10 independent components, whereas the Lanczos tensor has 20 independent components. Thus equation (4.30) is an overdetermined system. This freedom, however, can be removed by introducing two gauge conditions.

Lanczos showed that the left-hand side of (4.25) remains invariant under the maps

$$\begin{aligned} H_{abc} &\longrightarrow H_{abc} - V_b g_{ac} + V_a g_{bc} \\ Q_{ab} &\longrightarrow Q_{ab} + V_{a;\,b} + V_{b;\,a} - \frac{1}{2} V^d{}_{;\,d} g_{ab} \\ q &\longrightarrow q - \frac{1}{2} V^d{}_{;\,d} \end{aligned} \tag{4.31}$$

where V_a is an arbitrary vector. By choosing

$$V_a = -\tfrac{1}{2} H_{abc} g^{bc}$$

he obtained what is now known as the trace-free gauge condition

$$H_a{}^b{}_b = 0. \tag{4.32}$$

This reduces the number of independent components of H_{abc} from 20 to 16. A further reduction can be made by introducing what has come to be known as the divergence-free gauge condition

$$H_{ab}{}^c{}_{;c} = 0 \tag{4.33}$$

which now reduces the number of independent components of H_{abc} from 16 to 10, thus keeping in line with the number of independent components of the Weyl tensor. Although the trace-free gauge condition applied to (4.28) and (4.30) will reduce the number of terms in these equations, the divergence-free condition will not affect them directly. However, by covariantly differentiating (4.11),

$$H_{abc;\,d} + H_{bca;\,d} + H_{cab;\,d} = 0,$$

and contracting with g^{cd} yields

$$H_{ab}{}^{c}{}_{;c} + H_{b}{}^{c}{}_{a;c} + H^{c}{}_{ab;c} = 0.$$

Comparing this with the divergence–free condition gives

$$H_{a}{}^{c}{}_{b;c} = H_{b}{}^{c}{}_{a;c}. \tag{4.34}$$

Hence, substituting (4.32) and (4.34) into (4.28) and (4.30) yields the following simplified equations

$$\begin{aligned} R_{abcd} = {} & H_{abc;d} + H_{cda;b} + H_{bad;c} + H_{dcb;a} \\ & + H^{e}{}_{ac;e} g_{bd} + H^{e}{}_{bd;e} g_{ac} \\ & - H^{e}{}_{ad;e} g_{bc} - H^{e}{}_{bc;e} g_{ad} \\ & + \frac{1}{6} R (g_{ac} g_{bd} - g_{ad} g_{bc}), \end{aligned} \tag{4.35}$$

and

$$\begin{aligned} C_{abcd} = {} & H_{abc;d} + H_{cda;b} + H_{bad;c} + H_{dcb;a} \\ & + H^{e}{}_{ac;e} g_{bd} + H^{e}{}_{bd;e} g_{ac} \\ & - H^{e}{}_{ad;e} g_{bc} - H^{e}{}_{bc;e} g_{ad}. \end{aligned} \tag{4.36}$$

At the time of writing equation (4.25) Lanczos was unaware that the tensor H_{abc} and its derivatives generated the Weyl tensor. However, following along similar lines to Rainich and Einstein, Lanczos decided to split the Riemann tensor into two tensors which have 10 components each

$$U_{abcd} = A_{abcd} + \frac{1}{6} R (g_{ac} g_{bd} - g_{ad} g_{bc}) \tag{4.37}$$

$$V_{abcd} = S_{abcd} - \frac{1}{6} R (g_{ac} g_{bd} - g_{ad} g_{bc}) \tag{4.38}$$

where

$$R_{abcd} = \frac{1}{2} (U_{abcd} + V_{abcd}).$$

From this splitting Lanczos produced a new relation analogous to (4.4); namely,

$$\begin{aligned} U_{abcd} = {} & (R_{ac} - \frac{1}{6} R g_{ac}) g_{bd} + (R_{bd} - \frac{1}{6} R g_{bd}) g_{ac} \\ & - (R_{ad} - \frac{1}{6} R g_{ad}) g_{bc} - (R_{bc} - \frac{1}{6} R g_{bc}) g_{ad}. \end{aligned} \tag{4.39}$$

He then subtracted this from his version of (4.35) where he defined $R_{ac} = R_{abc}{}^{b}$. The resulting equation gives (4.36) with the left-hand side as $\frac{1}{2}V_{abcd}$. It was Takeno (Takeno, 1964) who first showed that the tensor V_{abcd} is proportional to the Weyl tensor. Clearly, in the same way as $(R_{ac} - \frac{1}{6}Rg_{ac})$ generates U_{abcd} algebraically, H_{abc} generates C_{abcd} differentially.

One may ask if it is possible to reduce H_{abc} such that itself can be generated by more fundamental elements. Locally, the Lanczos tensor cannot be reduced to the metric tensor unless we consider linearised gravitational theory close to flat space-time. Then we can write the metric tensor in terms of the flat space-time tensor η_{ab} and a perturbation quantity h_{ab}

$$g_{ab} = \eta_{ab} + h_{ab}. \tag{4.40}$$

Using (4.39) Lanczos expressed H_{abc} locally as

$$H_{abc} = \frac{1}{24}(6h_{ac,b} - 6h_{bc,a} + 2h_{a}{}^{d}{}_{,d}\,\eta_{bc} - 2h_{b}{}^{d}{}_{,d}\,\eta_{ac}).$$

One also may ask whether it would be possible to construct for the Riemann tensor an analogous expression to (4.36): the *Riemann–Lanczos* problem, as it is known. It has in fact been argued (Udeschini 1977, 1978, 1980, 1981) that the Riemann tensor can be generated by the derivatives of the Lanczos tensor alone

$$R_{abcd} = H_{abc;d} + H_{cda;b} + H_{bad;c} + H_{dcb;a}. \tag{4.41}$$

This is just equation (4.35) without the summed terms and the term involving the Ricci scalar. However, it has been shown (Massa and Pagani, 1984; Edgar, 1987) that the proposition (4.41) is false in general relativity. But these proofs relied on certain assumptions of dimension and signature. Later, a new proof was given (Edgar, 1994) demonstrating that (4.41) is invalid for all dimensions equal to or greater than four and for any signature.

It was sometime before a proof was given that confirmed the existence of the Lanczos tensor for any space-time (Bampi and Caviglia, 1983); indeed, this fact was missing from Lanczos' original paper. To this end an analogous problem was considered which necessarily omitted the cyclic condition (4.11). The proof was eventually extended (Illge, 1988) to confirm the existence of the Lanczos *spinor* for the *Weyl-Lanczos* problem.

4.4 The Lanczos spinor

We will now utilise the decomposition techniques first encountered in Sec. 2.8 and apply them to the Lanczos tensor.

Firstly, let us write the skew-symmetry properties that the Lanczos tensor exhibits on its first two indices,

$$H_{abc} = -H_{bac}, \tag{4.42}$$

in spinor form

$$H_{AA'BB'CC'} = -H_{BB'AA'CC'}. \tag{4.43}$$

Decomposing (4.43) yields

$$\begin{aligned} H_{abc} &= H_{ABA'B'CC'} \\ &= \frac{1}{2}(H_{ABA'B'CC'} - H_{ABB'A'CC'}) \\ &\quad + \frac{1}{2}(H_{ABB'A'CC'} - H_{BAB'A'CC'}). \end{aligned} \tag{4.44}$$

The first parenthesis is skew-symmetric in A', B', implying, from (4.43), that it is symmetric in A, B. The second parenthesis is skew-symmetric in A, B implying, again from (4.43), that it must also be symmetric in A', B'. Hence

$$H_{abc} = H_{ABA'B'CC'} = H_{(AB)[A'B']CC'} + H_{[AB](B'A')CC'}. \tag{4.45}$$

Recall that for a general multivalent spinor $K\ldots_{DE}\ldots$ (see (2.57)) we have

$$K\ldots_{[AB]}\ldots = \frac{1}{2}\epsilon_{AB}\, K\ldots_{D}{}^{D}\ldots. \tag{4.46}$$

Thus writing (4.45) in terms of (4.46) gives

$$\begin{aligned} H_{abc} = H_{ABA'B'CC'} &= \frac{1}{2}\epsilon_{A'B'}\, H_{(AB)D'}{}^{D'}{}_{CC'} \\ &\quad + \frac{1}{2}\epsilon_{AB}\, H_{D}{}^{D}{}_{(B'A')CC'}. \end{aligned} \tag{4.47}$$

Let us define the following spinors by

$$H_{ABCC'} = H_{(AB)CC'} = \frac{1}{2}H_{(AB)D'}{}^{D'}{}_{CC'}$$

and

$$\phi_{A'B'CC'} = \phi_{(A'B')CC'} = \frac{1}{2}H_{D}{}^{D}{}_{(A'B')CC'}. \tag{4.48}$$

We can then write (4.47) as

$$H_{abc} = H_{ABA'B'CC'} = \epsilon_{A'B'} H_{ABCC'} + \epsilon_{AB} \phi_{A'B'CC'}. \tag{4.49}$$

We chose $H_{ABCC'}$ and $\phi_{A'B'CC'}$ purposely to be symmetric in A, B and A', B' respectively in order that H_{abc} in (4.42) remains skew-symmetric in a, b. Conjugating (4.49) yields

$$H_{abc} = \overline{H}_{A'B'ABC'C} = \epsilon_{AB} \overline{H}_{A'B'C'C} + \epsilon_{A'B'} \overline{\phi}_{ABC'C}. \tag{4.50}$$

Because H_{abc} is real, and by comparing (4.49) with (4.50), we get

$$\begin{aligned} H_{abc} &= \overline{H}_{A'B'ABC'C} = H_{ABA'B'CC'} \\ &= \epsilon_{A'B'} H_{ABCC'} + \epsilon_{AB} \phi_{A'B'CC'}. \end{aligned}$$

Transvecting this with ϵ^{AB} gives

$$\phi_{A'B'CC'} = \frac{1}{2} \overline{H}_{A'B'D}{}^{D}{}_{C'C} = \overline{H}_{A'B'C'C},$$

the last term was obtained by employing (4.48). Substituting this into (4.49) yields

$$H_{abc} = \epsilon_{A'B'} H_{ABCC'} + \epsilon_{AB} \overline{H}_{A'B'C'C}. \tag{4.51}$$

Up to this point we have not imposed the cyclic condition (4.11), nor the two gauge conditions, (4.32) and (4.33), which means that the only symmetry properties that the Lanczos spinor $H_{ABCC'}$ possesses is

$$H_{ABCC'} = H_{(AB)CC'}. \tag{4.52}$$

Let us now obtain the spinor version of Lanczos' trace-free gauge condition (4.32),

$$H_a{}^b{}_b = H_{abc} g^{bc} = H_{AA'BB'CC'} \epsilon^{BC} \epsilon^{B'C'}.$$

Using (1.50) this becomes

$$\begin{aligned} H_a{}^b{}_b &= (H_{ABCC'} \epsilon_{A'B'} + \overline{H}_{A'B'C'C} \epsilon_{AB}) \epsilon^{BC} \epsilon^{B'C'} \\ &= -H_{ABCA'} \epsilon^{BC} - \overline{H}_{A'B'C'A} \epsilon^{B'C'} \\ &= 0 \end{aligned}$$

or

$$H_{ABCA'} \epsilon^{BC} = -\overline{H}_{A'B'C'A} \epsilon^{B'C'}. \tag{4.53}$$

Previously, we introduced condition (4.10) to reduce the number of components of the Lanczos tensor from 24 to 20. The first term in (4.10) can be written in spinor form as

$$\begin{aligned}{}^*H_{abc}g^{bc} &= (-iH_{ABCC'}\,\epsilon_{A'B'} + i\overline{H}_{A'B'C'C}\epsilon_{AB})\,\epsilon^{BC}\epsilon^{B'C'}\\ &= 0\end{aligned}$$

or

$$H_{ABCA'}\,\epsilon^{BC} = \overline{H}_{A'B'C'A}\,\epsilon^{B'C'}. \tag{4.54}$$

Thus comparing (4.53) with (4.54) implies that

$$H_{AD}{}^{D}{}_{A'} = 0. \tag{4.55}$$

From (4.46) we can write (4.55) as

$$\frac{1}{2}\epsilon_{BC}H_{AD}{}^{D}{}_{A'} = H_{A[BC]A'} = 0$$

which implies that

$$H_{ABCC'} = H_{A(BC)C'}. \tag{4.56}$$

Comparing (4.52) with (4.56) gives the Lanczos spinor with all its symmetry properties

$$H_{ABCC'} = H_{(ABC)C'}. \tag{4.57}$$

Clearly, (4.57) possesses eight independent complex components.

4.5 The spinor version of the Weyl–Lanczos equations

From above we can now write the corresponding spinor version of equation (4.36). We obtain

$$\begin{aligned}C_{abcd} &= \Psi_{ABCD}\,\epsilon_{A'B'}\,\epsilon_{C'D'} + \overline{\Psi}_{A'B'C'D'}\,\epsilon_{AB}\,\epsilon_{CD}\\ &= \nabla_{DD'}(H_{ABCC'}\,\epsilon_{A'B'} + \overline{H}_{A'B'C'C}\,\epsilon_{AB})\\ &\quad -\nabla_{CC'}(H_{ABDD'}\,\epsilon_{A'B'} + \overline{H}_{A'B'D'D}\,\epsilon_{AB})\\ &\quad +\nabla_{BB'}(H_{CDAA'}\,\epsilon_{C'D'} + \overline{H}_{C'D'A'A}\,\epsilon_{CD})\\ &\quad -\nabla_{AA'}(H_{CDBB'}\,\epsilon_{C'D'} + \overline{H}_{C'D'B'B}\,\epsilon_{CD})\\ &\quad -\nabla_{EE'}(H_{A}{}^{E}{}_{CC'}\,\epsilon_{A'}{}^{E'} + \overline{H}_{A'}{}^{E'}{}_{C'C}\,\epsilon_{A}{}^{E})\epsilon_{BD}\epsilon_{B'D'}\\ &\quad -\nabla_{EE'}(H_{B}{}^{E}{}_{DD'}\,\epsilon_{B'}{}^{E'} + \overline{H}_{B'}{}^{E'}{}_{D'D}\,\epsilon_{B}{}^{E})\epsilon_{AC}\epsilon_{A'C'}\\ &\quad +\nabla_{EE'}(H_{A}{}^{E}{}_{DD'}\,\epsilon_{A'}{}^{E'} + \overline{H}_{A'}{}^{E'}{}_{D'D}\,\epsilon_{A}{}^{E})\epsilon_{BC}\,\epsilon_{B'C'}\end{aligned}$$

$$+\nabla_{EE'}(H_B{}^E{}_{CC'}\epsilon_{B'}{}^{E'} + \overline{H}_{B'}{}^{E'}{}_{C'C}\,\epsilon_B{}^E)\epsilon_{AD}\epsilon_{A'D'}.$$

Transvecting this with $\epsilon^{A'B'}\epsilon^{C'D'}$ and then simplifying yields

$$\begin{aligned} 2\Psi_{ABCD} &= \nabla_D{}^{E'}H_{ABCE'} + \nabla_C{}^{E'}H_{ABDE'} \\ &\quad +\nabla_B{}^{E'}H_{CDAE'} + \nabla_A{}^{E'}H_{CDBE'}. \end{aligned} \tag{4.58}$$

So far we have not used the divergence-free gauge condition (4.33). In the spinor version this becomes

$$\begin{aligned} H_{ab}{}^c{}_{;c} &= \nabla_{CC'}(H_{AB}{}^{CC'}\epsilon_{A'B'} + \overline{H}_{A'B'}{}^{C'C}\epsilon_{AB}) \\ &= 0. \end{aligned}$$

As the covariant derivative of the metric tensor vanishes we have

$$\nabla_{CC'}H_{AB}{}^{CC'}\epsilon_{A'B'} + \nabla_{CC'}\overline{H}_{A'B'}{}^{C'C}\epsilon_{AB} = 0. \tag{4.59}$$

Transvecting with $\epsilon^{A'B'}$ gives

$$\nabla^{CC'}H_{ABCC'} = 0. \tag{4.60}$$

It is sometimes usual for authors to leave the expression for the Weyl–Lanczos equations as they stand in (4.58) and (4.60). However, they can be written in a slightly more compact way. If we employ (4.46) then (4.60) can be written as

$$\frac{1}{2}\epsilon_{DE}\nabla^{CC'}H_{ABCC'} = \nabla_{[E}{}^{C'}H_{|AB|D]C'} = 0$$

or

$$\nabla_E{}^{C'}H_{ABDC'} = \nabla_D{}^{C'}H_{ABEC'}. \tag{4.61}$$

Applying (4.61) to (4.58) gives

$$\Psi_{ABCD} = \nabla_D{}^{E'}H_{ABCE'} + \nabla_B{}^{E'}{}_{ADCE'}.$$

A second application yields

$$\Psi_{ABCD} = 2\nabla_D{}^{E'}H_{ABCE'} \tag{4.62}$$

which is the final spinor version of the Weyl–Lanczos equations with both gauge conditions imposed.

4.6 The Lanczos coefficients

The symmetry properties exhibited by the Lanczos spinor in (4.57) can be used to obtain eight complex scalar quantities in terms of the spin basis o^A, ι^A. These eight quantities are defined as follows:

$$\begin{aligned}
H_0 &= H_{0000'} = H_{ABCC'}o^A o^B o^C o^{C'} \\
H_1 &= H_{0010'} = H_{ABCC'}o^A o^B \iota^C o^{C'} \\
H_2 &= H_{0110'} = H_{ABCC'}o^A \iota^B \iota^C o^{C'} \\
H_3 &= H_{1110'} = H_{ABCC'}\iota^A \iota^B \iota^C o^{C'} \\
H_4 &= H_{0001'} = H_{ABCC'}o^A o^B o^C \iota^{C'} \\
H_5 &= H_{0011'} = H_{ABCC'}o^A o^B \iota^C \iota^{C'} \\
H_6 &= H_{0111'} = H_{ABCC'}o^A \iota^B \iota^C \iota^{C'} \\
H_7 &= H_{1111'} = H_{ABCC'}\iota^A \iota^B \iota^C \iota^{C'}.
\end{aligned} \tag{4.63}$$

Each one of the relations in (4.63) can be expressed in terms of the null tetrad vectors **l**, **n**, **m** and $\overline{\mathbf{m}}$ in a straightforward way:

$$\begin{aligned}
H_0 &= H_{abc}l^a m^b l^c = H_{131} \\
H_1 &= H_{abc}l^a m^b \overline{m}^c = H_{134} \\
H_2 &= H_{abc}\overline{m}^a n^b l^c = H_{421} \\
H_3 &= H_{abc}\overline{m}^a n^b \overline{m}^c = H_{424} \\
H_4 &= H_{abc}l^a m^b m^c = H_{133} \\
H_5 &= H_{abc}l^a m^b n^c = H_{132} \\
H_6 &= H_{abc}\overline{m}^a n^b m^c = H_{423} \\
H_7 &= H_{abc}\overline{m}^a n^b n^c = H_{422}.
\end{aligned} \tag{4.64}$$

To show that (4.63) and (4.64) are equivalent let us attempt to derive, for example, the H_5 term in (4.63) from the corresponding term in (4.64). Starting with the expression (4.51) our task is to transvect through by the appropriate combination of spin basis components such that the conjugated term on the right-hand side of (4.51) vanishes, but leaving the correct form of H_5 as given in (4.63). With the aid of (2.95) one can see straight away that the only combination will be

$$l^a m^b m^c = o^A o^{A'} o^B \iota^{B'} \iota^C \iota^{C'}.$$

Thus

$$H_{abc}l^a m^b n^c = \epsilon_{A'B'} H_{ABCC'}o^A o^{A'} o^B \iota^{B'} \iota^C \iota^{C'}$$

and the result follows.

4.7 The Weyl–Lanczos equations in spin coefficient form

When trying to find solutions of the Weyl–Lanczos equations (4.62) it is more convenient to transform them into relations involving the Lanczos coefficients and spin coefficients.

Many solutions of the Einstein field equations have been found by specifying a particular choice of null tetrad. This choice automatically defines the intrinsic derivatives (3.116), and using (3.128) will result in four identities involving spin coefficients and metric functions. If the scalar ϕ is replaced with the coordinates of the space-time concerned in each of the four identities of (3.128), 16 algebraic equations involving the, as yet, unknown spin coefficients will be obtained. Solving these equations will give the spin coefficients in terms of the metric functions — but not necessarily explicitly. The Newman–Penrose field equations can then be employed to yield expressions for the Weyl scalars, again, not necessarily explicitly.

It is for those space-times which have known spin coefficients and Weyl scalars that the Weyl–Lanczos equations can be solved, and expressions for the Lanczos coefficients obtained.

Let us start by writing the dyad components of the Weyl spinor with respect to an orthonormal basis

$$\Psi_{\mathbf{ABCD}} = \Psi_{ABCD}\epsilon_{\mathbf{A}}{}^{A}\epsilon_{\mathbf{B}}{}^{B}\epsilon_{\mathbf{C}}{}^{C}\epsilon_{\mathbf{D}}{}^{D}. \tag{4.65}$$

Substituting (4.62) directly into (4.65) gives

$$\Psi_{\mathbf{ABCD}} = (2\nabla_{D}{}^{E'}H_{ABCE'})\epsilon_{\mathbf{A}}{}^{A}\epsilon_{\mathbf{B}}{}^{B}\epsilon_{\mathbf{C}}{}^{C}\epsilon_{\mathbf{D}}{}^{D},$$

which can be written as

$$\Psi_{\mathbf{ABCD}} = (2\nabla_{D}{}^{E'}H_{\mathbf{PQRS'}}\epsilon_{A}{}^{\mathbf{P}}\epsilon_{B}{}^{\mathbf{Q}}\epsilon_{C}{}^{\mathbf{R}}\epsilon_{E'}{}^{\mathbf{S'}})\epsilon_{\mathbf{A}}{}^{A}\epsilon_{\mathbf{B}}{}^{B}\epsilon_{\mathbf{C}}{}^{C}\epsilon_{\mathbf{D}}{}^{D}. \tag{4.66}$$

Applying Liebniz's law (3.5) to each quantity within the parentheses of (4.66) yields

$$\begin{aligned}\Psi_{\mathbf{ABCD}} = 2(&\epsilon^{\mathbf{S'X'}}\nabla_{\mathbf{DX'}}H_{\mathbf{ABCS'}} + \epsilon^{\mathbf{S'X'}}H_{\mathbf{PBCS'}}\epsilon_{\mathbf{A}}{}^{A}\nabla_{\mathbf{DX'}}\epsilon_{A}{}^{\mathbf{P}}\\ &+\epsilon^{\mathbf{S'X'}}H_{\mathbf{AQCS'}}\epsilon_{\mathbf{B}}{}^{B}\nabla_{\mathbf{DX'}}\epsilon_{B}{}^{\mathbf{Q}} + \epsilon^{\mathbf{S'X'}}H_{\mathbf{ABRS'}}\epsilon_{\mathbf{C}}{}^{C}\nabla_{\mathbf{DX'}}\epsilon_{C}{}^{\mathbf{R}}\\ &+\epsilon^{\mathbf{Y'W'}}H_{\mathbf{ABCS'}}\epsilon_{\mathbf{Y'}}{}^{E'}\nabla_{\mathbf{W'D}}\epsilon_{E'}{}^{\mathbf{S'}}),\end{aligned} \tag{4.67}$$

then employing the relations defining the Ricci rotation coefficients (3.105) and (3.106) we obtain

$$\begin{aligned}\Psi_{\mathbf{ABCD}} = 2(&\epsilon^{\mathbf{S'X'}}\nabla_{\mathbf{DX'}}H_{\mathbf{ABCS'}} - \epsilon^{\mathbf{S'X'}}H_{\mathbf{PBCS'}}\gamma_{\mathbf{DX'A}}{}^{\mathbf{P}} \\ &-\epsilon^{\mathbf{S'X'}}H_{\mathbf{AQCS'}}\gamma_{\mathbf{DX'B}}{}^{\mathbf{Q}} - \epsilon^{\mathbf{S'X'}}H_{\mathbf{ABRS'}}\gamma_{\mathbf{DX'C}}{}^{\mathbf{R}} \\ &-\epsilon^{\mathbf{Y'W'}}H_{\mathbf{ABCS'}}\overline{\gamma}_{\mathbf{W'DY'}}{}^{\mathbf{S'}}). \end{aligned} \tag{4.68}$$

It is now a straightforward task to replace the dyad indices with the appropriate numerical values 0, 1. Thence referring to (3.114), (3.135) and (4.63) we have the following form of the Weyl-Lanczos equations, (see also Maher and Zund, 1968; Zund, 1975)

$$\begin{aligned}
\Psi_0 &= -2[DH_4 - \delta H_0 \\
&\quad +(\overline{\epsilon} - 3\epsilon - \overline{\rho})\,H_4 + (\overline{\alpha} + 3\beta - \overline{\pi})\,H_0 + 3\kappa\,H_5 - 3\sigma\,H_1] \\
\Psi_1 &= -2[DH_5 - \delta H_1 \\
&\quad +(\overline{\epsilon} - \epsilon - \overline{\rho})\,H_5 + (\overline{\alpha} + \beta - \overline{\pi})\,H_1 + 2\kappa\,H_6 - 2\sigma H_2 - \pi H_4 + \mu H_0] \\
\Psi_1 &= -2[\overline{\delta} H_4 - \Delta H_0 \\
&\quad +(\overline{\beta} - 3\alpha - \overline{\tau})\,H_4 + (-\overline{\mu} + 3\gamma + \overline{\gamma})\,H_0 + 3\rho H_5 - 3\tau H_1] \\
\Psi_2 &= -2[DH_6 - \delta H_2 \\
&\quad +(\overline{\epsilon} + \epsilon - \overline{\rho})\,H_6 + (\overline{\alpha} - \beta - \overline{\pi})\,H_2 - 2\pi H_5 + 2\mu H_1 + \kappa\,H_7 - \sigma H_3] \\
\Psi_2 &= -2[\overline{\delta} H_5 - \Delta H_1 \\
&\quad +(\overline{\beta} - \alpha - \overline{\tau})\,H_5 + (-\overline{\mu} + \gamma + \overline{\gamma})\,H_1 + 2\rho H_6 - 2\tau H_2 - \lambda\,H_4 + \nu H_0] \\
\Psi_3 &= -2[DH_7 - \delta H_3 \\
&\quad +(\overline{\epsilon} + 3\epsilon - \overline{\rho})\,H_7 + (\overline{\alpha} - 3\beta - \overline{\pi})\,H_3 - 3\pi H_6 + 3\mu H_2] \\
\Psi_3 &= -2[\overline{\delta} H_6 - \Delta H_2 \\
&\quad +(\overline{\beta} + \alpha - \overline{\tau})\,H_6 + (-\overline{\mu} - \gamma + \overline{\gamma})\,H_2 - 2\lambda\,H_5 + 2\nu H_1 + \rho H_7 - \tau H_3] \\
\Psi_4 &= -2[\overline{\delta} H_7 - \Delta H_3 \\
&\quad +(\overline{\beta} + 3\alpha - \overline{\tau})\,H_7 + (-\overline{\mu} - 3\gamma + \overline{\gamma})\,H_3 - 3\lambda\,H_6 + 3\nu\,H_2].
\end{aligned} \tag{4.69}$$

4.8 The Ricci–Lanczos equations in spin coefficient form

Let us now utilise the technique of the previous section and apply it to the Ricci–Lanczos equations (4.26).

Applying the conditions (4.32) and (4.34) to the expression for the Ricci–Lanczos equations (4.26) yields

$$R_{bc} = 2H_{abc;d}\,g^{ad} - 2Q_{bc} + 6qg_{bc}. \tag{4.70}$$

This can be written in spinor notation as

$$R_{BCB'C'} = 2[\nabla_{B'}{}^{A} H_{ABCC'} + \nabla_{B}{}^{A'} \overline{H}_{A'B'C'C} - \Theta_{BCB'C'} + 3q\epsilon_{BC}\epsilon_{B'C'}] \qquad (4.71)$$

where $\Theta_{BCB'C'} = \Theta_{(BC)(B'C')} = Q_{bc}$, and Q_{bc} is trace-free (see Sec. 2.8.1). Comparing (4.71) with the relation involving the Ricci spinor (3.40) we obtain

$$\Phi_{BCB'C'} = -[\nabla_{B'}{}^{A} H_{ABCC'} + \nabla_{B}{}^{A'} \overline{H}_{A'B'C'C} - \Theta_{BCB'C'} + 3(q-\Lambda)\epsilon_{BC}\epsilon_{B'C'}]. \qquad (4.72)$$

On transvecting through by $\epsilon^{BC}\epsilon^{B'C'}$ one immediately finds that

$$q = \Lambda. \qquad (4.73)$$

Thus

$$\Phi_{BCB'C'} = -[\nabla_{B'}{}^{A} H_{ABCC'} + \nabla_{B}{}^{A'} \overline{H}_{A'B'C'C} - \Theta_{BCB'C'}]. \qquad (4.74)$$

The spinor dyad form of (4.74) can easily be found. Thus

$$\begin{aligned}\Phi_{\mathbf{BCB'C'}} &= \epsilon^{\mathbf{XP}} \nabla_{\mathbf{X\,B'}} H_{\mathbf{PBCC'}} + \epsilon^{\mathbf{XY}} H_{\mathbf{PBCC'}}\, \gamma_{\mathbf{X\,B'Y}}{}^{\mathbf{P}} \\ &\quad + \epsilon^{\mathbf{XP}} H_{\mathbf{PQCC'}}\, \gamma_{\mathbf{X\,B'B}}{}^{\mathbf{Q}} + \epsilon^{\mathbf{XP}} H_{\mathbf{PBRC'}}\, \gamma_{\mathbf{X\,B'C}}{}^{\mathbf{R}} \\ &\quad + \epsilon^{\mathbf{XP}} H_{\mathbf{PBCC'}}\, \overline{\gamma}_{\mathbf{B'\,XC'}}{}^{\mathbf{S'}} + c.c. \\ &\quad + \Theta_{\mathbf{BCB'C'}}\end{aligned} \qquad (4.75)$$

where *c.c.* stands for the complex conjugate of *all* the preceding terms.

Employing (3.116), (3.118), (3.136) and (4.63), the spin coefficient form of (4.75) can now be given by

$$\begin{aligned}\Phi_{00} &= D(H_1 + \overline{H}_1) - \delta\overline{H}_0 - \overline{\delta} H_0 \\ &\quad - H_1(3\rho + \epsilon + \overline{\epsilon}) - \overline{H}_1(3\overline{\rho} + \epsilon + \overline{\epsilon}) - H_4\overline{\sigma} - \overline{H}_4\sigma \\ &\quad - \overline{H}_0(\overline{\pi} - 3\overline{\alpha} - \beta) - H_0(\pi - 3\alpha - \overline{\beta}) \\ &\quad + (\overline{H}_2 + H_5)\overline{\kappa} - (H_2 + \overline{H}_5)\kappa - \Theta_{00} \\ \Phi_{11} &= \tfrac{1}{2}[D(H_6 + 2\overline{H}_6) - \delta(H_2 + 2\overline{H}_5) - \Delta H_1 - \overline{\delta} H_5 \\ &\quad - H_6(2\rho - \overline{\rho} - \overline{\epsilon} - \epsilon) - 2\overline{H}_6(2\overline{\rho} - \overline{\epsilon} - \epsilon) + H_3\sigma \\ &\quad - H_5(2\pi + \overline{\tau} - \overline{\alpha} + \beta) - 2\overline{H}_5(2\overline{\pi} - \overline{\alpha} + \beta) + H_0\nu \\ &\quad - H_2(2\tau + \overline{\alpha} - \beta) - \overline{H}_2\overline{\pi} + 2\overline{H}_4\overline{\lambda} + H_4\lambda \\ &\quad - H_1(2\mu - \overline{\mu} - \overline{\gamma} - \gamma) 2\overline{H}_1\mu + 2\overline{H}_7\overline{\kappa} + H_7\kappa + \Theta_{11}] \\ \Phi_{01} &= D(\overline{H}_2 + H_5) - \delta\overline{H}_1 + \overline{\delta} H_4\end{aligned}$$

$$
\begin{aligned}
&-\overline{H}_2(2\overline{\rho}-\overline{\epsilon}+\epsilon)-H_5(3\rho-\overline{\epsilon}+\epsilon)\\
&-\overline{H}_1(2\overline{\pi}-\overline{\alpha}-\beta)-H_4(\pi-3\alpha+\overline{\beta})\\
&-\overline{H}_3\overline{\kappa}+\overline{H}_6\kappa-H_0\overline{\mu}-\overline{H}_0\overline{\lambda}+\overline{H}_5\sigma-H_1\overline{\pi}+\Theta_{01}\\
\Phi_{12}=&-\Delta(\overline{H}_2+H_5)+\delta H_6+\overline{\delta}\,\overline{H}_3\\
&-\overline{H}_2(3\overline{\mu}-\gamma+\overline{\gamma})-H_5(2\mu-\gamma+\overline{\gamma})\\
&-\overline{H}_3(\overline{\tau}+\alpha-3\overline{\beta})-H_6(2\tau+\overline{\alpha}-3\beta)\\
&+(2\overline{H}_1+H_1)\overline{\nu}+H_4\nu+\overline{H}_7\rho+H_7\sigma-H_2\overline{\lambda}-\Theta_{12}\\
\Phi_{10}=&\,D(H_2+\overline{H}_5)-\overline{\delta}H_1+\delta\overline{H}_4\\
&-H_2(2\rho-\epsilon+\overline{\epsilon})-\overline{H}_5(3\overline{\rho}-\epsilon+\overline{\epsilon})\\
&-H_1(2\pi-\alpha-\overline{\beta})-\overline{H}_4(\overline{\pi}-3\overline{\alpha}+\beta)\\
&-H_3\kappa+H_6\overline{\kappa}-\overline{H}_0\mu-H_0\lambda+H_5\overline{\sigma}-\overline{H}_1\pi+\Theta_{10}\\
\Phi_{21}=&-\Delta(H_2+\overline{H}_5)+\overline{\delta}\,\overline{H}_6+\delta H_3\\
&-H_2(3\mu-\overline{\gamma}+\gamma)-\overline{H}_5(2\overline{\mu}-\overline{\gamma}+\gamma)\\
&-H_3(\tau+\overline{\alpha}-3\beta)-\overline{H}_6(2\overline{\tau}+\alpha-3\overline{\beta})\\
&+(2H_1+\overline{H}_1)\nu+\overline{H}_4\overline{\nu}+H_7\overline{\rho}+\overline{H}_7\overline{\sigma}-\overline{H}_2\lambda-\Theta_{21}\\
\Phi_{02}=&\,D\overline{H}_3-\Delta H_4+\delta(H_5-\overline{H}_2)\\
&-\overline{H}_3(\overline{\rho}-3\overline{\epsilon}+\epsilon)-H_4(\mu+\overline{\gamma}-3\gamma)\\
&-H_5(3\tau-\overline{\alpha}+\beta)-\overline{H}_2(3\overline{\pi}+\overline{\alpha}-\beta)\\
&-(H_1-2\overline{H}_1)\overline{\lambda}+(2H_6-\overline{H}_6)\sigma\\
&+H_0\overline{\nu}+\overline{H}_7\kappa-\Theta_{02}\\
\Phi_{22}=&-\Delta(\overline{H}_6+H_6)+\delta H_7+\overline{\delta}\,\overline{H}_7\\
&-3\overline{H}_6\overline{\mu}-3H_6\mu-H_7(\tau-\overline{\alpha}-3\beta)-\overline{H}_7(\overline{\tau}-\alpha-3\overline{\beta})\\
&+(\overline{H}_2+2H_5)\nu+(H_2+2\overline{H}_5)\overline{\nu}\\
&-\overline{H}_3\lambda-H_3\overline{\lambda}-\Theta_{22}\\
\Phi_{20}=&\,DH_3-\Delta\overline{H}_4+\overline{\delta}(\overline{H}_5-H_2)\\
&-H_3(\rho-3\epsilon+\overline{\epsilon})-\overline{H}_4(\overline{\mu}+\gamma-3\overline{\gamma})\\
&-\overline{H}_5(3\overline{\tau}-\alpha+\overline{\beta})-H_2(3\pi+\alpha-\overline{\beta})\\
&-(\overline{H}_1-2H_1)\lambda+(2\overline{H}_6-H_6)\overline{\sigma}\\
&+\overline{H}_0\nu+H_7\overline{\kappa}-\Theta_{20},
\end{aligned}
\tag{4.76}
$$

where we define the dyad components of the spinor $\Theta_{\mathbf{ABC'D'}}$ by

$$
\begin{aligned}
\Theta_{00}&=\Theta_{000'0'}=\Theta_{ABC'D'}o^Ao^Bo^{C'}o^{D'}\\
\Theta_{10}&=\Theta_{010'0'}=\Theta_{ABC'D'}o^A\iota^Bo^{C'}o^{D'}
\end{aligned}
$$

$$\begin{aligned}
\Theta_{20} &= \Theta_{110'0'} = \Theta_{ABC'D'}\iota^A\iota^B o^{C'} o^{D'}\\
\Theta_{01} &= \Theta_{000'1'} = \Theta_{ABC'D'}o^A o^B o^{C'} \iota^{D'}\\
\Theta_{11} &= \Theta_{010'1'} = \Theta_{ABC'D'}o^A\iota^B o^{C'} \iota^{D'}\\
\Theta_{21} &= \Theta_{110'1'} = \Theta_{ABC'D'}\iota^A\iota^B o^{C'} \iota^{D'}\\
\Theta_{02} &= \Theta_{001'1'} = \Theta_{ABC'D'}o^A o^B \iota^{C'} \iota^{D'}\\
\Theta_{12} &= \Theta_{011'1'} = \Theta_{ABC'D'}o^A\iota^B \iota^{C'} \iota^{D'}\\
\Theta_{22} &= \Theta_{111'1'} = \Theta_{ABC'D'}\iota^A\iota^B \iota^{C'} \iota^{D'}.
\end{aligned} \tag{4.77}$$

Also, in terms of null tetrad vectors:

$$\begin{aligned}
\Theta_{00} &= Q_{ab}l^a l^b = Q_{(1)(1)}\\
\Theta_{01} &= Q_{ab}l^a m^b = Q_{(1)(3)}\\
\Theta_{02} &= Q_{ab}m^a m^b = Q_{(3)(3)}\\
\Theta_{10} &= Q_{ab}l^a \overline{m}^b = Q_{(1)(4)}\\
\Theta_{11} &= Q_{ab}l^a n^a + Q_{ab}m^a \overline{m}^a\\
&= Q_{(1)(2)} + Q_{(3)(4)}\\
\Theta_{12} &= Q_{ab}m^a n^b = Q_{(3)(2)}\\
\Theta_{20} &= Q_{ab}\overline{m}^a \overline{m}^b = Q_{(4)(4)}\\
\Theta_{21} &= Q_{ab}\overline{m}^a n^b = Q_{(4)(2)}\\
\Theta_{22} &= Q_{ab}n^a n^b = Q_{(2)(2)}.
\end{aligned} \tag{4.78}$$

Clearly, it can be seen from (4.76) that for empty space-times $\Theta_{\mathbf{ABC'D'}}$ is expressed by Lanczos coefficients and spin coefficients alone.

4.9 The behaviour of Lanczos coefficients under Lorentz transformations

We will carry out a similar analysis to that carried out in Sec. 3.6. Again, we impose the orthonormalisation conditions (2.96) on the 'old' and 'new' tetrads of (3.142), leading to three rotations and a Lorentz boost. Each of these transformations leaves (2.96), or equivalently (2.94), invariant.

4.9.1 *Null rotation with* l *fixed*

In this case the null tetrad transforms as

$$l^a \mapsto l^a$$

$$\begin{aligned} n^a &\mapsto n^a + \bar{c}m^a + c\overline{m}^a + c\bar{c}l^a \\ m^a &\mapsto m^a + cl^a \\ \overline{m}^a &\mapsto \overline{m}^a + \bar{c}l^a \end{aligned} \tag{4.79}$$

where $c = c_2$ of (3.142).

The spin basis transforms as

$$\begin{aligned} o^A &\mapsto o^A \\ \iota^A &\mapsto \iota^A + \bar{c}o^A. \end{aligned} \tag{4.80}$$

The Lanczos coefficients transform as

$$\begin{aligned} H_0 &\mapsto H_0 \\ H_1 &\mapsto H_1 + 2\bar{c}H_0 \\ H_2 &\mapsto H_2 + \bar{c}^2 H_0 + 2\bar{c}H_1 \\ H_3 &\mapsto H_3 + \bar{c}^3 H_0 + 3\bar{c}^2 H_1 + 3\bar{c}H_2 \\ H_4 &\mapsto H_4 + cH_0 \\ H_5 &\mapsto H_5 + c\bar{c}H_0 + cH_1 + \bar{c}H_4 \\ H_6 &\mapsto H_6 + c\bar{c}^2 H_0 + 2c\bar{c}H_1 + cH_2 + \bar{c}^2 H_4 + 2\bar{c}H_5 \\ H_7 &\mapsto H_7 + c\bar{c}^3 H_0 + 3c\bar{c}^2 H_1 + 3c\bar{c}H_2 + cH_3 + \bar{c}^3 H_4 + 3\bar{c}^2 H_5 + 3\bar{c}H_6. \end{aligned} \tag{4.81}$$

4.9.2 *Null rotation with* n *fixed*

The null tetrad transforms as

$$\begin{aligned} l^a &\mapsto l^a + \bar{c}m^a + c\overline{m}^a + c\bar{c}n^a \\ n^a &\mapsto n^a \\ m^a &\mapsto m^a + cn^a \\ \overline{m}^a &\mapsto \overline{m}^a + \bar{c}n^a \end{aligned} \tag{4.82}$$

where $c = c_1$ of (3.142).

The spin basis transforms as

$$\begin{aligned} o^A &\mapsto o^A + c\iota^A \\ \iota^A &\mapsto \iota^A. \end{aligned} \tag{4.83}$$

The Lanczos coefficients transform as

$$H_0 \mapsto H_0 + 3cH_1 + 3c^2 H_2 + c^3 H_3 + \bar{c}H_4 + 3c\bar{c}H_5 + 3c^2\bar{c}H_6 + c^3\bar{c}H_7$$

$$
\begin{aligned}
H_1 &\mapsto H_1 + 2cH_2 + c^2H_3 + \bar{c}H_5 + 2c\bar{c}H_6 + c^2\bar{c}H_7 \\
H_2 &\mapsto H_2 + cH_3 + \bar{c}H_6 + c\bar{c}H_7 \\
H_3 &\mapsto H_3 + \bar{c}H_7 \\
H_4 &\mapsto H_4 + 3cH_5 + 3c^2H_6 + c^3H_7 \\
H_5 &\mapsto H_5 + 2cH_6 + c^2H_7 \\
H_6 &\mapsto H_6 + cH_7 \\
H_7 &\mapsto H_7.
\end{aligned}
\tag{4.84}
$$

4.9.3 *Spin-boost transformation*

The null tetrad transforms as

$$
\begin{aligned}
l^a &\mapsto al^a \\
n^a &\mapsto \frac{1}{a}n^a \\
m^a &\mapsto e^{i\theta}m^a \\
\overline{m}^a &\mapsto e^{-i\theta}\overline{m}^a
\end{aligned}
\tag{4.85}
$$

where $a = a_1 = 1/a_4$ and $e^{i\theta} = c_5$ of (3.142).

The spin basis transforms as

$$
\begin{aligned}
o^A &\mapsto a^{1/2}e^{i\theta/2}o^A \\
\iota^A &\mapsto a^{-1/2}e^{i\theta/2}\iota^A.
\end{aligned}
\tag{4.86}
$$

The real spatial rotation parameter θ and the boost parameter a are defined in Sec. 3.6.3.

The Lanczos coefficients transform as

$$
\begin{aligned}
H_0 &\mapsto a^2e^{i\theta}H_0 \\
H_1 &\mapsto aH_1 \\
H_2 &\mapsto e^{-i\theta}H_2 \\
H_3 &\mapsto \frac{1}{a}e^{-2i\theta}H_3 \\
H_4 &\mapsto ae^{2i\theta}H_4 \\
H_5 &\mapsto e^{i\theta}H_5 \\
H_6 &\mapsto \frac{1}{a}H_6 \\
H_7 &\mapsto \frac{1}{a^2}e^{-i\theta}H_7.
\end{aligned}
\tag{4.87}
$$

4.10 Miscellaneous transformations

4.10.1 $\mathrm{l} \leftrightarrow \mathrm{n}$ *and* $\mathrm{m} \leftrightarrow -\overline{\mathrm{m}}$

The null tetrad transforms as

$$\begin{aligned} l^a &\mapsto n^a \\ n^a &\mapsto l^a \\ m^a &\mapsto -\overline{m}^a \\ \overline{m}^a &\mapsto -m^a. \end{aligned} \tag{4.88}$$

The spin basis transforms as

$$\begin{aligned} o^A &\mapsto \iota^A \\ \iota^A &\mapsto -o^A. \end{aligned} \tag{4.89}$$

The Lanczos coefficients transform as

$$\begin{aligned} H_0 &\mapsto H_7 \\ H_1 &\mapsto -H_6 \\ H_2 &\mapsto H_5 \\ H_3 &\mapsto -H_4 \\ H_4 &\mapsto -H_3 \\ H_5 &\mapsto H_2 \\ H_6 &\mapsto -H_1 \\ H_7 &\mapsto H_0. \end{aligned} \tag{4.90}$$

4.10.2 *Prime operation*

The null tetrad transforms as

$$\begin{aligned} (l^a)' &= n^a \\ (n^a)' &= l^a \\ (m^a)' &= \overline{m}^a \\ (\overline{m}^a)' &= m^a. \end{aligned} \tag{4.91}$$

The spin basis transforms as

$$\begin{aligned} (o^A)' &= i\iota^A \\ (\iota^A)' &= io^A \end{aligned}$$

$$
\begin{aligned}
(o^{A'})' &= -i\iota^{A'} \\
(\iota^{A'})' &= -io^{A'}.
\end{aligned} \tag{4.92}
$$

The Lanczos coefficients transform as

$$
\begin{aligned}
H_0' &= H_7 \\
H_1' &= H_6 \\
H_2' &= H_5 \\
H_3' &= H_4 \\
H_4' &= -H_3 \\
H_5' &= -H_2 \\
H_6' &= -H_1 \\
H_7' &= -H_0.
\end{aligned} \tag{4.93}
$$

4.10.3 *Asterisk operation*

The null tetrad transforms as

$$
\begin{aligned}
(l^a)^* &= m^a \\
(n^a)^* &= -\overline{m}^a \\
(m^a)^* &= -l^a \\
(\overline{m}^a)^* &= n^a.
\end{aligned} \tag{4.94}
$$

The spin basis transforms as

$$
\begin{aligned}
(o^A)^* &= o^A \\
(\iota^A)^* &= \iota^A \\
(o^{A'})^* &= \iota^{A'} \\
(\iota^{A'})^* &= -o^{A'}.
\end{aligned} \tag{4.95}
$$

The Lanczos coefficients transform as

$$
\begin{aligned}
H_0^* &= H_4 \\
H_1^* &= H_5 \\
H_2^* &= H_6 \\
H_3^* &= H_7 \\
H_4^* &= -H_0 \\
H_5^* &= -H_1
\end{aligned}
$$

$$\begin{aligned} H_6^* &= -H_2 \\ H_7^* &= -H_3. \end{aligned} \tag{4.96}$$

4.11 The Weyl–Lanczos equations in GHP form

The Weyl–Lanczos equations display a particularly elegant form when written in the GHP formalism. Namely

$$\begin{aligned} \Psi_0 &= 2[\eth H_0 + \text{þ} H_3' - \bar{\tau}' H_0 - \bar{\rho} H_3' + 3\sigma H_1 + 3\kappa H_2'] \\ \Psi_1 &= 2[\text{þ}' H_0 + \eth' H_3' - \bar{\rho}' H_0 - \bar{\tau} H_3' + 3\tau H_1 + 3\rho H_2'] \\ \Psi_1 &= 2[\eth H_1 + \text{þ} H_2' + \rho' H_0 - \bar{\tau}' H_1 + 2\sigma H_2 + \tau' H_3' - \bar{\rho} H_2' + 2\kappa H_1'] \\ \Psi_2 &= 2[\text{þ}' H_1 + \eth' H_2' - \bar{\rho}' H_1 + \kappa' H_0 + 2\tau H_2 + \sigma' H_3' - \bar{\tau}' H_2 + 2\rho H_1'] \\ \Psi_2 &= 2[\eth H_2 + \text{þ} H_1' - \bar{\tau}' H_2 + 2\rho' H_1 + \sigma H_3 + 2\tau' H_2' - \bar{\rho} H_1' + \kappa' H_0] \\ \Psi_3 &= 2[\text{þ}' H_2 + \eth' H_1' - \bar{\rho}' H_2 + 2\kappa' H_1 + \tau H_3 + 2\sigma' H_2' - \bar{\tau} H_1' + \rho H_0'] \\ \Psi_3 &= 2[\eth H_3 + \text{þ} H_0' - \bar{\tau}' H_3 + 3\rho' H_2 + 3\tau' H_1' - \bar{\rho} H_0'] \\ \Psi_4 &= 2[\text{þ}' H_3 + \eth' H_0' - \bar{\rho}' H_3 + 3\kappa' H_2 + 3\sigma' H_1' - \bar{\tau} H_0]. \end{aligned} \tag{4.97}$$

All the Lanczos coefficients transform according to (3.184) and hence classify as weighted scalars. The types of these coefficients are given by

$$\begin{aligned} H_0 &: \{3,1\} - \text{scalar} \\ H_0' &: \{3,1\} - \text{scalar} \\ H_1 &: \{1,1\} - \text{scalar} \\ H_1' &: \{1,1\} - \text{scalar} \\ H_2 &: \{-1,1\} - \text{scalar} \\ H_2' &: \{-1,1\} - \text{scalar} \\ H_3 &: \{-3,1\} - \text{scalar} \\ H_3' &: \{-3,1\} - \text{scalar}. \end{aligned} \tag{4.98}$$

4.12 Solutions of the Weyl–Lanczos equations

Explicit forms of the Lanczos tensor for a given space-time are notoriously difficult to obtain. We have already discussed that with the aid of the NP field equations, Bianchi identities, etc. expressions for the spin coefficients are readily obtainable. Unfortunately, there are no general methods that one can utilise to yield analogous expressions for the Lanczos coefficients. It is perhaps for this reason alone that the Lanczos tensor has been neglected for many years, as an analysis of its qualitative role would be exceedingly difficult without analytic forms.

It is not my intention to give a complete and detailed account of all the solutions of the Weyl–Lanczos equations, even though the number of metrics for which explicit forms exist is comparatively small. The reason for this is because many of the explicit forms of the Lanczos coefficients have been obtained via a purely heuristic approach.

A number of different approaches have been analysed in the quest to arrive at explicit forms. For example, a kinematical method was developed which yielded forms for the Schwarzschild, Kasner and Gödel metrics (see Novello and Velloso, 1987). In that particular case the Weyl–Lanczos equations (4.28) were considered. However, most progress to date has been made when the spin coefficient form of the Weyl–Lanczos equations (4.69) has been analysed. Consequently, we will look at some solutions of (4.69) for vacuum space-times, which will be discovered to be directly expressible mainly in terms of spin coefficients. Note that the forms of the Lanczos coefficients are by no means unique.

4.12.1 *Petrov type O space-times*

By choosing an appropriate form of the null tetrad (2.95) and referring to Table 2.9.3 we see that this Petrov type describes Minkowski space. Consequently the dyad components of the Weyl spinor Ψ_i $(i = 1, 2, 3, 4, 5)$ all vanish. Comparing the Weyl–Lanczos equations (4.69) with the NP field equations (3.132) suggests that a possible solution of (4.69) is the trivial one:

$$\begin{aligned} H_0 &= 0 \\ H_1 &= 0 \\ H_2 &= 0 \\ H_3 &= 0 \\ H_4 &= 0 \\ H_5 &= 0 \\ H_6 &= 0 \\ H_7 &= 0. \end{aligned} \tag{4.99}$$

4.12.2 Petrov type N space-times

We can choose the null tetrad such that $\Psi_i = 0,\ i \neq 4$. Then comparing the Weyl–Lanczos equations (4.69) with the NP field equations (3.132) suggests that a possible solution of (4.69) is

$$\begin{aligned} H_0 &= \frac{\kappa}{2} \\ H_1 &= \frac{\rho}{6} \\ H_2 &= -\frac{\pi}{6} \\ H_3 &= -\frac{\lambda}{2} \\ H_4 &= \frac{\sigma}{2} \\ H_5 &= \frac{\tau}{6} \\ H_6 &= -\frac{\mu}{6} \\ H_7 &= -\frac{\nu}{2}. \end{aligned} \tag{4.100}$$

4.12.3 Petrov type III space-times

Choosing the null tetrad such that $\Psi_i = 0,\ i \neq 3$ and comparing the Weyl–Lanczos equations (4.69) with the NP field equations (3.132) suggests that a possible solution of (4.69) is

$$\begin{aligned} H_0 &= \kappa \\ H_1 &= \frac{\rho}{3} \\ H_2 &= -\frac{\pi}{3} \\ H_3 &= -\lambda \\ H_4 &= \sigma \\ H_5 &= \frac{\tau}{3} \\ H_6 &= -\frac{\mu}{3} \\ H_7 &= -\nu. \end{aligned} \tag{4.101}$$

Notice that the Lanczos coefficients in (4.100) differ from those in (4.101) by a multiple of two. Furthermore, each Lanczos coefficient is represented by a single spin coefficient.

4.12.4 *Lanczos coefficients for the Schwarzschild metric*

There is evidence to suggest (O'Donnell, 1997) that for Petrov type D space-times the following conditions on the Lanczos coefficients hold:

$$H_0 = H_4 \tag{4.102}$$

$$H_1 = H_5 \tag{4.103}$$

$$H_2 = H_6 \tag{4.104}$$

$$H_3 = H_7. \tag{4.105}$$

Let us postulate these conditions for the Schwarzschild metric

$$ds^2 = -(1-2m/r)^{-1}dr^2 - r^2(d\theta^2 + \sin^2\theta d\phi^2) + (1-2m/r)dt^2, \tag{4.106}$$

which is Petrov type D.

It is rather a simple procedure to obtain the explicit form of the null tetrad for the metric (4.102), (Kinnersley, 1969). We have

$$\begin{aligned} l^a &= \left(1-\frac{2m}{r}\right)^{-1} \delta^a{}_0 + \delta^a{}_1 \\ n^a &= \frac{1}{2}\delta^a{}_0 - \frac{1}{2}r\left(1-\frac{2m}{r}\right)\delta^a{}_1 \\ m^a &= \frac{1}{\sqrt{2}\,r}(\delta^a{}_2 - i\csc\theta\delta^a{}_3). \end{aligned}$$

However, a more useful form of this null tetrad, for our purposes, can be obtained by applying a Lorentz boost transformation (3.149), yielding

$$\begin{aligned} l^a &= \frac{1}{\sqrt{2}}\left(1-\frac{2m}{r}\right)^{-\frac{1}{2}} \delta^a{}_4 + \frac{1}{\sqrt{2}}\left(1-\frac{2m}{r}\right)^{\frac{1}{2}} \delta^a{}_1 \\ n^a &= \frac{1}{\sqrt{2}}\left(1-\frac{2m}{r}\right)^{-\frac{1}{2}} \delta^a{}_4 - \frac{1}{\sqrt{2}}\left(1-\frac{2m}{r}\right)^{\frac{1}{2}} \delta^a{}_1 \\ m^a &= \frac{1}{\sqrt{2}\,r}(\delta^\alpha{}_2 - i\csc\theta\delta^a{}_3), \end{aligned} \tag{4.107}$$

with corresponding intrinsic derivatives given by

$$\begin{aligned} D &= \frac{1}{\sqrt{2}}\left[\left(1-\frac{2m}{r}\right)^{-\frac{1}{2}}\frac{\partial}{\partial t} + \left(1-\frac{2m}{r}\right)^{\frac{1}{2}}\frac{\partial}{\partial r}\right] \\ \Delta &= \frac{1}{\sqrt{2}}\left[\left(1-\frac{2m}{r}\right)^{-\frac{1}{2}}\frac{\partial}{\partial t} - \left(1-\frac{2m}{r}\right)^{\frac{1}{2}}\frac{\partial}{\partial r}\right] \end{aligned}$$

$$\delta = \frac{1}{\sqrt{2}r}\left[\frac{\partial}{\partial\theta} - i\csc\theta\,\frac{\partial}{\partial\phi}\right]. \tag{4.108}$$

All type D metrics are algebraically special (see Sec. 2.92). Choosing l^a and n^a as principal null directions implies that $\Psi_0 = \Psi_1 = \Psi_3 = \Psi_4 = 0$. So from the Goldberg–Sachs theorem $\kappa = \sigma = 0$; and from Exercise 3.13, $\nu = \lambda = 0$. For the particular choice of null tetrad (4.107) we have

$$\Psi_2 = -\frac{m}{r^3}, \tag{4.109}$$

and the only non-vanishing spin coefficients are

$$\begin{aligned}\rho = \mu &= -\frac{1}{\sqrt{2}\,r}\left(1 - \frac{2m}{r}\right)^{\frac{1}{2}}\\ \gamma = \epsilon &= \frac{m}{2\sqrt{2}\,r^2}\left(1 - \frac{2m}{r}\right)^{-\frac{1}{2}}\\ \alpha = -\beta &= -\frac{\cot\theta}{2\sqrt{2}r}.\end{aligned} \tag{4.110}$$

It is reasonable to suggest that any Lanczos coefficient found by solving (4.69) with the conditions (4.109) and (4.110) would be, at most, dependent on r and θ. This is by virtue of the fact that neither Ψ_2 nor the non-zero spin coefficients have t, ϕ dependence. We will assume, therefore, that $H_i = H_i(r, \theta)$ for $i = 0, 1, 2$ which means that the intrinsic derivatives are simplified somewhat, i.e. $\Delta = -D$ and $\bar{\delta} = \delta$. Therefore together with (4.102)–(4.105), and the non-vanishing spin coefficients and Weyl scalars, we can rewrite the Weyl–Lanczos equations (4.69) as

$$DH_3 - \delta H_0 - (2\epsilon + \rho)H_3 - 2\alpha H_0 = 0 \tag{4.111}$$

$$DH_2 - \delta H_1 - \rho H_2 + \rho H_0 = 0 \tag{4.112}$$

$$DH_0 + \delta H_3 - 4\alpha H_3 + (4\epsilon - \rho)H_0 + 3\rho H_2 = 0 \tag{4.113}$$

$$DH_1 - \delta H_2 + (2\epsilon + \rho)H_1 + 2\alpha H_2 = -\frac{\Psi_2}{2} \tag{4.114}$$

$$DH_1 + \delta H_2 + (2\epsilon + \rho)H_1 - 2\alpha H_2 = -\frac{\Psi_2}{2} \tag{4.115}$$

$$DH_0 - \delta H_3 + 4\alpha H_3 + (4\epsilon - \rho)H_0 + 3\rho H_2 = 0 \tag{4.116}$$

$$DH_2 + \delta H_1 - \rho H_2 + \rho H_0 = 0 \tag{4.117}$$

$$DH_3 + \delta H_0 - (2\epsilon + \rho)H_3 + 2\alpha H_0 = 0. \tag{4.118}$$

These can be further simplified by summing and subtracting (4.111),

(4.118); (4.112), (4.117); (4.113), (4.116) and (4.114), (4.115):

$$DH_3 = (2\epsilon + \rho)H_3 \tag{4.119}$$

$$\delta H_0 = -2\alpha H_0 \tag{4.120}$$

$$DH_2 = \rho H_2 - \rho H_0 \tag{4.121}$$

$$\delta H_1 = 0 \tag{4.122}$$

$$DH_0 = -(4\epsilon - \rho)H_0 - 3\rho H_2 \tag{4.123}$$

$$\delta H_3 = 4\alpha H_3 \tag{4.124}$$

$$DH_1 = -(2\epsilon + \rho)H_1 - \frac{\Psi_2}{2} \tag{4.125}$$

$$\delta H_2 = 2\alpha H_2. \tag{4.126}$$

Solving (4.119) and (4.124) for H_3 yields

$$H_3 = \frac{1}{r}\left(1 - \frac{2m}{r}\right)^{\frac{1}{2}} \csc^2\theta. \tag{4.127}$$

Taking mixed derivatives of (4.121) and (4.126), then subtracting the two expressions gives

$$(\delta D - D\delta)H_2 = \delta(\rho H_2) - \delta(\rho H_0) - 2D(\alpha H_2).$$

Also, from the commutation relations (3.128)(2) with $\phi = H_2$, we have

$$(\delta D - D\delta)H_2 = \rho\delta H_2.$$

Comparing these two expressions with each other results in

$$H_0 = 0, \tag{4.128}$$

and from (4.123),

$$H_2 = 0. \tag{4.129}$$

Clearly, from (4.122), H_1 is independent of θ. So solving (4.125) for H_1 yields

$$H_1 = -\frac{m}{3\sqrt{2}r^2}\left(1 - \frac{2m}{r}\right)^{-\frac{1}{2}}. \tag{4.130}$$

It is possible now to express (4.127) and (4.130) mainly in terms of spin coefficients. In conclusion then, the full compliment of Lanczos coefficients is, with the aid of (4.110), given by

$$\begin{aligned} H_0 &= H_4 = 0 \\ H_1 &= H_5 = 0 \\ H_2 &= H_6 = -\frac{2}{3}\epsilon \\ H_3 &= H_7 = -\sqrt{2}\,\rho \csc^2\theta. \end{aligned}$$

4.13 A brief note on the Lanczos spinor/tensor

There are indications that the work Lanczos conducted in his 1938 paper on variational principles which are quadratic in the components of the Riemann tensor (Lanczos, 1938) indirectly led to the discovery of his 'spin-tensor' in 1962. Although this tensor first appeared in his 1949 paper, the results therein were restricted to weak fields (Lanczos, 1949). The vast amount of work pertaining to the Lanczos spinor/tensor (especially since 1990) would fill a number of large volumes. Consequently this chapter was intended to give only a flavour of this research, and is not a review. The bibliography contains many important and significant contributions on the Lanczos spinor/tensor, but it is by no means an exhaustive survey. The interested reader should consult these references for appropriate articles in order to comprehensively study this area.

4.14 Exercises

4.1 Show that if the trace-free gauge condition (4.32) and the divergence-free condition (4.33) are not imposed then the Weyl–Lanczos equations become

$$\Psi_{ABCD} = 2\nabla_D{}^{E'} H_{ABCE'}.$$

4.2 Assume that in addition to abandoning the gauge conditions (4.32) and (4.33), the cyclic condition (4.11) is also removed. Show that the spin coefficient form of the Weyl–Lanczos equations is given by

$$\begin{aligned} \Psi_0 = &-2[DH_4 - \delta H_0 - (\overline{\rho} + 3\epsilon - \overline{\epsilon})H_4 + (H_1 + H_8 - H_{11})\sigma \\ &+(-\overline{\pi} + \overline{\alpha} + 3\beta)H_0 - (H_5 + H_{10} - H_{13})\kappa] \\ \Psi_1 = &-\frac{1}{2}[D(H_5 + H_{10} - H_{13}) - \delta(H_1 + H_8 - H_{11}) + \overline{\delta}H_4 - \Delta H_0 \\ &+(H_5 + H_{10} - H_{13})(\rho - \overline{\rho} - \epsilon + \overline{\epsilon}) \end{aligned}$$

$$
\begin{aligned}
&+(H_1+H_8-H_{11})\,(\overline{\alpha}+\beta-\tau-\overline{\pi})\\
&+(\overline{\beta}-3\alpha-\overline{\tau}-3\pi)H_4+(\overline{\alpha}+3\alpha-\overline{\mu}+\mu)H_0\\
&+2\sigma(H_2-\overline{H}_{10}-\overline{H}_{13})+2\kappa(H_6+H_9-H_{12})]\\
\Psi_2 =& -\frac{1}{3}[D(3H_8-2H_9+2H_6-\overline{H}_6)-\delta(\overline{H}_5+2H_2+\overline{H}_{13})\\
&+\overline{\delta}(2H_5+\overline{H}_2-H_{13})-\Delta(3H_9-2H_8+2H_1-\overline{H}_1)\\
&+H_8(\overline{\gamma}+\gamma-\overline{\mu}+2\mu)+H_9(\overline{\epsilon}+\epsilon-\overline{\rho}+2\rho)\\
&-H_1(\overline{\gamma}+\gamma-\mu+2\overline{\mu})-H_6(\overline{\epsilon}+\epsilon-\rho+2\overline{\rho})\\
&-\overline{H}_1(\overline{\gamma}+\gamma-\mu-\overline{\mu})-\overline{H}_6(\overline{\epsilon}+\epsilon+\rho-\overline{\rho})\\
&-H_5(\pi+2\overline{\tau}+2\alpha+2\overline{\beta})-\overline{H}_5(\overline{\pi}-\tau-\overline{\alpha}+\beta)\\
&+\overline{H}_2(\pi-\overline{\tau}-\alpha+\overline{\beta})+H_2(2\overline{\pi}-\tau+2\overline{\alpha}+2\beta)\\
&+H_{13}(2\pi+\overline{\tau}+\alpha-\overline{\beta})-\overline{H}_{13}(\overline{\pi}+2\tau-\overline{\alpha}+\beta)]\\
\Psi_3 =& -\frac{1}{2}[\delta(\overline{H}_6+H_9+H_{12})-\Delta(\overline{H}_2+H_{10}+H_{13})\ +D\overline{H}_7-\overline{\delta H}_3\\
&+(\overline{H}_6+H_9+H_{12})\,(\overline{\alpha}+\beta-\tau-\overline{\pi})\\
&+(\overline{H}_2+H_{10}+H_{13})\,(\overline{\mu}-\mu+\gamma-\overline{\gamma})\\
&+\overline{H}_7(\epsilon+3\overline{\epsilon}-\rho+3\overline{\rho})+\overline{H}_3(\alpha-3\overline{\beta}-\pi-3\overline{\tau})\\
&+2\overline{\nu}(\overline{H}_1+H_8+H_{11})-2\overline{\lambda}(\overline{H}_5+\overline{H}_{10}-\overline{H}_{13})]\\
\Psi_4 =& -2[\overline{\delta}H_7-\Delta H_3+(-\overline{\tau}+\overline{\beta}+3\alpha)H_7+(H_2+\overline{H}_{10}+\overline{H}_{13})\nu\\
&-(\overline{\mu}+3\gamma-\overline{\gamma})H_3-(H_6+H_9-H_{12})\lambda],
\end{aligned}
$$

where $H_0,\ldots,H_7$ are the Lanczos coefficients given in (4.63) and (4.64), and $H_8,\ldots,H_{13}$ are given by

$$
\begin{aligned}
H_8 &= H_{121} = H_{abc}\,l^a n^b l^c\\
H_9 &= H_{122} = H_{abc}\,l^a n^b n^c\\
H_{10} &= H_{123} = H_{abc}\,l^a n^b m^c\\
H_{11} &= H_{341} = H_{abc}\,m^a \overline{m}^b l^c\\
H_{12} &= H_{342} = H_{abc}\,m^a \overline{m}^b n^c\\
H_{13} &= H_{343} = H_{abc}\,m^a \overline{m}^b m^c.
\end{aligned}
$$

4.3 The following *extended* exercise is based on the paper "The connection between general observers and the Lanczos potential" (Novello and Velloso, 1987).
Consider an arbitrary timelike vector field $V_a(x^a)$, $V^aV_a = 1$, then the

covariant derivative of this vector field can be decomposed:

$$V_{a;b} = \sigma_{ab} + \omega_{ab} + (\theta/3)\, h_{ab} + a_a V_b \tag{4.131}$$

where in this case V_a is the 4-velocity of a perfect fluid which has energy momentum tensor

$$T_{ab} = \mu V_a V_b - p h_{ab}. \tag{4.132}$$

The quantities in (4.131) and (4.132) are defined as follows.
The projection tensor

$$h_{ab} = g_{ab} - V_a V_b \tag{4.133}$$

is a map $T \to S$, where T is a tangent space and S is the space of all vectors orthogonal to V_a i.e. the three-dimensional rest space. The quantities

$$a_a = V_{a;b} V^b \tag{4.134}$$

$$\theta = V^a{}_{;a} \tag{4.135}$$

$$\omega_{ab} = h_{[a}{}^e\, h_{b]}{}^f V_{e;f} \tag{4.136}$$

$$\theta_{ab} = h_{(a}{}^e\, h_{b)}{}^f V_{e;f} - (\theta/3) h_{ab} \tag{4.137}$$

are known respectively as the acceleration, expansion, rotation, and shear. The quantities μ and p are the energy–density and pressure of the fluid respectively. The above quantities obey the following conditions:

$$\sigma_{ab} V^b = 0 \tag{4.138}$$
$$\omega_{ab} V^b = 0 \tag{4.139}$$
$$a_a V^a = 0 \tag{4.140}$$
$$h_{ab} V^b = 0 \tag{4.141}$$
$$\sigma_{ab} g^{ab} = 0 \tag{4.142}$$
$$h_{ab} g^{ab} = 3 \tag{4.143}$$
$$h_a{}^c h_{cb} = h_{ab} \tag{4.144}$$
$$\sigma_{ab} = \sigma_{(ab)} \tag{4.145}$$
$$\omega_{ab} = \omega_{[ab]} \tag{4.146}$$
$$h_{[ab]} = 0. \tag{4.147}$$

The Weyl tensor can be written with respect to the above quantities as

$$C_{abcd} = -2\{4V_{[a}E_{b][d}V_{c]} + g_{a[c}E_{d]b} - g_{b[c}E_{d]a} \\ -\epsilon_{abef}V^eV_{[c}H_{d]}{}^f - \epsilon_{cdef}V^eV_{[a}H_{b]}{}^f\} \tag{4.148}$$

where the 'electric' and 'magnetic' parts of the Weyl tensor are defined, respectively, by

$$-E_{ab} = h_a{}^e h_b{}^f \sigma_{ef;p}V^p - \frac{1}{3}h_{ab}(\omega_{pq}\omega^{pq} + \sigma_{pq}\sigma^{pq} - a^p{}_{;p}) \\ +a_a a_b - h_a{}^e h_b{}^f a_{(e;f)} + \frac{2}{3}\theta\sigma_{ab} + \sigma_a{}^e\sigma_{eb} - \omega_a\omega_b \tag{4.149}$$

and

$$-H_{ab} = h^e{}_{(a}h^f{}_{b)}\epsilon_e{}^{pq\sigma}V_\sigma(\omega_{fp} + \sigma_{fp})_{;q} - \frac{1}{2}a_{(a}\omega_{b)} \tag{4.150}$$

with

$$\omega^a = \frac{1}{2}\epsilon^{efqa}\omega_{ef}V_q.$$

By writing the null vectors l^a and n^a as linear combinations of the 4-velocity V^a and 4-acceleration a^a (see (2.108)) we obtain

$$l^a = \frac{1}{\sqrt{2}}\left(V^a - \frac{a^a}{\sqrt{-a_a\, a^a}}\right) \tag{4.151}$$

$$n^a = \frac{1}{\sqrt{2}}\left(V^a + \frac{a^a}{\sqrt{-a_a\, a^a}}\right). \tag{4.152}$$

Thence (see(2.109))

$$V^a = \frac{1}{\sqrt{2}}(l^a + n^a) \tag{4.153}$$

$$a^a = \frac{-a}{2}(l^a - n^a) \tag{4.154}$$

where $a^2 = -a_a\, a^a$. Due to the presence of the coefficient a in (4.154) we will instead use (4.134) as the defining equation for the 4-acceleration.

(a) Show that in terms of null vectors and spin coefficients:

$$a_a = \frac{1}{2}[(\epsilon + \overline{\epsilon} + \gamma + \overline{\gamma})(l_a - n_a) - (\overline{\kappa} - \nu + \overline{\tau} - \pi)(m_a + \overline{m}_a)] \tag{4.155}$$

$$h_{ab} = \frac{1}{2}[(l_a - n_a)(l_b - n_b) + 4m_{(a}\overline{m}_{b)}] \tag{4.156}$$

$$\theta = \frac{1}{\sqrt{2}}[(\epsilon + \overline{\epsilon}) - (\gamma + \overline{\gamma}) - (\rho - \overline{\mu}) - (\overline{\rho} - \mu)] \tag{4.157}$$

$$\begin{aligned}\omega_{ab} = \frac{1}{4\sqrt{2}}\{&[-\overline{\kappa} + \pi + \overline{\tau} - \nu - 2(\alpha + \overline{\beta})][(l_a - n_a)m_b + m_a(l_b - n_b)]\\ &+[-\kappa + \overline{\pi} + \tau - \overline{\nu} - 2(\overline{\alpha} + \beta)][(l_a - n_a)\overline{m}_b + \overline{m}_a(l_b - n_b)]\\ &+4(\rho - \overline{\mu} - \overline{\rho} + \mu)\overline{m}_{[a}m_{b]}\}\end{aligned} \tag{4.158}$$

and

$$\begin{aligned}\sigma_{ab} = \frac{1}{4\sqrt{2}}\Bigg\{&\frac{2}{3}[2\gamma + 2\overline{\gamma} - 2\epsilon - 2\overline{\epsilon} - \rho + \overline{\mu} - \overline{\rho} + \mu]\\ &\times[(l_a - n_a)(l_b - n_b) - 2m_{(a}\overline{m}_{b)}]\\ &+[\overline{\kappa} - \pi - \overline{\tau} + \nu - 2(\overline{\alpha} + \beta)][(l_a - n_a)m_b + m_a(l_b - n_b)]\\ &+[\kappa - \overline{\pi} - \tau + \overline{\nu} - 2(\alpha + \overline{\beta})][(l_a - n_a)\overline{m}_b + \overline{m}_a(l_b - n_b)]\\ &+4(\sigma - \overline{\lambda})\overline{m}_a\overline{m}_b + 4(\overline{\sigma} - \lambda)m_a m_b\Bigg\}.\end{aligned} \tag{4.159}$$

(b) Obtain similar expressions for the 'electric' and 'magnetic' parts of the Weyl tensor.

(c) The metric for the Kasner geometry (Kasner, 1921) is given by

$$ds^2 = -t^{2p_1}\,dx^2 - t^{2p_2}dy^2 - t^{2p_3}dz^2 + dt^2$$

where

$$p_1 + p_2 + p_3 = 1 \text{ and } (p_1)^2 + (p_2)^2 + (p_3)^2 = 1$$

p_1, p_2 and p_3 being real constants. A Lanczos tensor for this geometry was found to be

$$H_{abc} = \frac{1}{3}(\sigma_{ac}V_b - \sigma_{bc}V_a)$$

with $a_a = \omega_{ab} = 0$. (One can easily verify that this expression for the Lanczos tensor is correct. (4.36) and (4.148) should yield identical results with appropriate forms of (4.149) and (4.150) for this space-time.) For a particular choice of null tetrad the spin coefficients are given by

$$\gamma = -\epsilon = -\frac{p_1}{2\sqrt{2}t}$$

$$\rho = -\mu = -\frac{(p_3 + p_2)}{2\sqrt{2}t}$$

$$\lambda = -\sigma = -\frac{(p_3 - p_2)}{2\sqrt{2}t}.$$

Show that the only non-vanishing Lanczos coefficients are

$$H_1 = -H_6 = -\frac{1}{9}\left[\frac{3p_1 - 1}{2\sqrt{2}t}\right]$$

$$H_3 = -H_4 = \frac{1}{3}\left[\frac{p_3 - p_2}{2\sqrt{2}t}\right].$$

Appendix A

Aspects of general relativity

A.1 The space-time of general relativity

A.1.1 *Space-time as a differentiable manifold*

The mathematical model for physical space-time on which the special theory of relativity is based is affine Minkowski space M: that is, the underlying geometry of the theory is Minkowski geometry. The special theory is in excellent agreement with experiment — provided that gravitational phenomena are *not* present. Einstein's theory of gravitation is called the general theory of relativity and the mathematical model of physical space-time on which this theory is based is no longer affine Minkowski space M.

Our knowledge of the physical world is gained through observations of *events* (an event is something which occurs somewhere at sometime) and it seems reasonable to suppose that each event can be characterised by four coordinates (x^0, x^1, x^2, x^3), i.e. by four ordered real numbers, a member of the set $\mathbb{R}^4$. Furthermore it is reasonable to suppose that each coordinate can take values lying in some prescribed range, each set of values corresponding to some possible event (somewhere at sometime) and also to suppose that different events correspond to different values of the coordinates. It is too restrictive to suppose that one coordinate system can be taken to cover the set M of all possible events; instead, we suppose that M is the union of several subsets $U_1, U_2, \ldots$ (not necessarily disjoint) each of which can be covered by a single coordinate system. If two of the subsets U_α intersect then each event lying in the intersection will be characterised by two different coordinates. This is a situation with which we are familiar, i.e. when dealing with a transformation of coordinates (for example, Cartesians to polars). We then assume that the 'new' coordinates are differentiable functions of the 'old' and that the Jacobian is non-zero so that the coordinate transformation is invertible.

Of course, the set M of all possible events (i.e. all possible locations at all possible times) is usually referred to as physical space-time, and the observations made above can be formalised by stating that *M is supposed to form a four-dimensional differentiable manifold.* The remainder of the section will be devoted to recapitulating the definition of a differentiable manifold.

Definition 1 The set M is called an n-dimensional manifold if and only if $M = \bigcup_\alpha U_\alpha$ where, for each α, there exists a one-to-one mapping ψ_α from U_α on to an open subset of $\mathbb{R}^n$. If $P \in U_\alpha$ and

$$\psi_\alpha : P \rightarrow (x^1, x^2, \ldots, x^n)$$

then the multiple $(x^1, x^2, \ldots, x^n)$ is referred to as the *coordinates of P*. The subset U_α is called a *coordinate neighbourhood* of M and ψ_α is called a *coordinate function.* The pair (U_α, ψ_α) is called a *chart* or a (local) *coordinate system about P. P* itself is called a *point* of the manifold.

Definition 2 Since the maps ψ_α are one-to-one they are invertible. Whenever two coordinate neighbourhoods intersect, say U_a and U_b, the mapping $\psi_a \psi_b^{-1}$ will define a function, the range and image of the function being the coordinates of the points P lying in $U_a \bigcap U_b$ as evaluated in the charts (U_b, ψ_α) respectively. The manifold M is defined to be a *differentiable manifold* if these functions $\psi_a \psi_b^{-1}$, whenever they exist, are differentiable.

A.1.2 The tangent space $T_{P_O}(M)$ at a point P_O of a manifold M

Consider some surface in three-dimensional Euclidean space. Let O be an origin, then any point in the Euclidean space can be characterised by its position vector $\mathbf{r}$ relative to the chosen origin O. The characteristic of a *surface* is that the position vector of any point P on the surface can be written (at least locally) as a function of two parameters which we will denote here by x^1 and x^2:

$$\mathbf{r} = \mathbf{r}(x^1, x^2).$$

Such a surface is a two-dimensional manifold, the parameters x^1 and x^2 being the coordinates of the point P of the manifold. If x^2 is kept constant, say equal to c, then the equation

$$\mathbf{r} = \mathbf{r}(x^1, c)$$

will describe a curve lying on the surface with parameter x^1. A whole family of curves is obtained by letting c take different constant values. This family is called the family of x^1-parametric curves. A second family is obtained if x^1 is kept constant. Then,

$$\mathbf{r} = \mathbf{r}(c, x^2).$$

Through any point P_o having coordinates there will pass one and only one member of each of these functions, namely the curves

$$\mathbf{r} = \mathbf{r}(x^1, x_0^2) \text{ and } \mathbf{r} = \mathbf{r}(x_0^1, x^2).$$

The tangent vectors to these two curves at the point P_o are simply the partial derivatives

$$\left.\frac{\partial \mathbf{r}}{\partial x^1}\right|_{P_o} \text{ and } \left.\frac{\partial \mathbf{r}}{\partial x^2}\right|_{P_o}. \tag{A.1}$$

The set of vectors spanned by these two tangent vectors, i.e. all the vectors of the form

$$\mathbf{v} = v^1 \left.\frac{\partial \mathbf{r}}{\partial x^1}\right|_{P_o} + v^2 \left.\frac{\partial \mathbf{r}}{\partial x^2}\right|_{P_o} \tag{A.2}$$

where v^1 and v^2 are real constants (scalars) form a two-dimensional vector space and can be represented by arrows at P_o lying in the tangent plane to the surface at P_o. For this reason the two-dimensional vector space is called the *tangent space to the surface at the point* P_o. The two vectors $\frac{\partial \mathbf{r}}{\partial x^1}|_{P_o}$ and $\frac{\partial \mathbf{r}}{\partial x^2}|_{P_o}$ are called the natural basis vectors for the tangent space associated with the parameters x^1 and x^2. If a different choice of parameters is made then denoting these 'new' parameters by $x^{1'}$ and $x^{2'}$ the natural basis vectors become $\frac{\partial \mathbf{r}}{\partial x^{1'}}|_{P_o}$ and $\frac{\partial \mathbf{r}}{\partial x^{2'}}|_{P_o}$. In terms of these basis vectors the vector $\mathbf{v}$ is written as

$$\mathbf{v} = v^{1'} \left.\frac{\partial \mathbf{r}}{\partial x^{1'}}\right|_{P_o} + v^{2'} \left.\frac{\partial \mathbf{r}}{\partial x^{2'}}\right|_{P_o}. \tag{A.3}$$

It is easy to relate the two sets of components (v^1, v^2) and $(v^{1'}, v^{2'})$. Using the chain rule

$$\frac{\partial \mathbf{r}}{\partial x^1} = \frac{\partial \mathbf{r}}{\partial x^{1'}}\frac{\partial x^{1'}}{\partial x^1} + \frac{\partial \mathbf{r}}{\partial x^{2'}}\frac{\partial x^{2'}}{\partial x^1}$$

$$\frac{\partial \mathbf{r}}{\partial x^2} = \frac{\partial \mathbf{r}}{\partial x^{1'}}\frac{\partial x^{1'}}{\partial x^2} + \frac{\partial \mathbf{r}}{\partial x^{2'}}\frac{\partial x^{2'}}{\partial x^2}.$$

Hence (A.2) can be rewritten as

$$\begin{aligned}\mathbf{v} &= v^1\left(\left.\frac{\partial \mathbf{r}}{\partial x^{1'}}\right|_{P_o}\left.\frac{\partial x^{1'}}{\partial x^1}\right|_{P_o} + \left.\frac{\partial \mathbf{r}}{\partial x^{2'}}\right|_{P_o}\left.\frac{\partial x^{2'}}{\partial x^1}\right|_{P_o}\right)\\ &\quad + v^2\left(\left.\frac{\partial \mathbf{r}}{\partial x^{1'}}\right|_{P_o}\left.\frac{\partial x^{1'}}{\partial x^2}\right|_{P_o} + \left.\frac{\partial \mathbf{r}}{\partial x^{2'}}\right|_{P_o}\left.\frac{\partial x^{2'}}{\partial x^2}\right|_{P_o}\right)\\ &= \left(\left.\frac{\partial x^{1'}}{\partial x^1}\right|_{P_o} v^1 + \left.\frac{\partial x^{1'}}{\partial x^2}\right|_{P_o} v^2\right)\left.\frac{\partial \mathbf{r}}{\partial x^{1'}}\right|_{P_o}\\ &\quad + \left(\left.\frac{\partial x^{2'}}{\partial x^1}\right|_{P_o} v^1 + \left.\frac{\partial x^{2'}}{\partial x^2}\right|_{P_o} v^2\right)\left.\frac{\partial \mathbf{r}}{\partial x^{2'}}\right|_{P_o}.\end{aligned}$$

Comparing this with (A.3) gives

$$v^{1'} = \left.\frac{\partial x^{1'}}{\partial x^1}\right|_{P_o} v^1 + \left.\frac{\partial x^{1'}}{\partial x^2}\right|_{P_o} v^2 \text{ and } v^{2'} = \left.\frac{\partial x^{2'}}{\partial x^1}\right|_{P_o} \nu^1 \left.\frac{\partial x^{2'}}{\partial x^2}\right|_{P_o} v^2. \tag{A.4}$$

Now consider a general two-dimensional differentiable manifold, one which cannot necessarily be viewed as a surface in three-dimensional Euclidean space. Consider a local chart or coordinate system (U, ψ) about a point P_o in which the coordinates of a general point $P \in U$ are denoted by (x^1, x^2), the coordinates of P_o itself being denoted by (x^1_o, x^2_o). One-dimensional submanifolds are called curves and clearly the set of all points P corresponding to some fixed value of x^2 is such a manifold. Thus two families of parametric curves can be defined for the manifold in much the same way as for the surface. Unfortunately, expressions such as (A.1) and (A.2) cannot be written down because for a general two-dimensional manifold there will be no analogue of the (three-dimensional) position vector $\mathbf{r}$. However, dropping $\mathbf{r}$ from (A.1) and (A.2) yields the differential operators

$$\left.\frac{\partial}{\partial x^1}\right|_{P_o} \text{ and } \left.\frac{\partial}{\partial x^2}\right|_{P_o} \tag{A.5}$$

and then the linear differential operators

$$v^1\left.\frac{\partial}{\partial x^1}\right|_{P_o} + v^2\left.\frac{\partial}{\partial x^2}\right|_{P_o} \tag{A.6}$$

where v^1 and v^2 are real constants form a two-dimensional vector space over the real numbers. By analogy to the tangent space to the surface, this vector space is called the *tangent space to the manifold* at the point P_o. The differential operators (A.5) are called the *natural basis vectors* for the tangent space associated with the coordinates (x^1, x^2) or with the chart (U, ψ). If we choose another chart (U', ψ') about P_o, that is another coordinate system $(x^{1'}, x^{2'})$, then the components $v^{1'}, v^{2'}$ of a vector relative to the natural basis $\frac{\partial}{\partial x^{1'}}, \frac{\partial}{\partial x^{2'}}$ will be related to the components v^1, v^2 of the vector relative to the natural basis $\frac{\partial}{\partial x^1}, \frac{\partial}{\partial x^2}$ by the equations (A.4). Notice that the partial derivatives $\frac{\partial x^{1'}}{\partial x^1}|_{P_o}$ etc. exist because the manifold is assumed to be differentiable.

Everything said above for a two-dimensional manifold can be generalised to an n-dimensional manifold and in particular to the four-dimensional space-time manifold M. Thus each point P_o of M there is defined a vector space, the tangent space to M at P_o. This vector space will be denoted by $T_{P_o}(M)$, it is defined over the real numbers, that is the scalars are the real numbers. It is convenient to introduce enumerative indices $i, j, k, l, \ldots$ which range from 0 to 3 so that the natural basis for the vector space $T_{P_o}(M)$ associated with the coordinate system $x^i (i = 0, 1, 2, 3)$ can be written as

$$\frac{\partial}{\partial x^i} \ (i = 0, 1, 2, 3)$$

and the formula corresponding to (A.4) for the transformation of the components of a given vector under a transformation of coordinates from $x^i (i = 0, 1, 2, 3)$ to $x^{i'} (i = 0, 1, 2, 3)$ can be written as

$$v^{i'} = \sum_{j=0}^{3} \frac{\partial x^{i'}}{\partial x^i}\bigg|_{P_o} v^j \ (i = 0, 1, 2, 3).$$

Of course this notation is simplified considerably by using the usual conventions that

(1) Single indices (so called *free* indices) range from 0 to 3.
(2) Repeated indices (i.e. indices repeated in any term or product of terms — so called *dummy* or *summation* indices) sum from 0 to 3.

With these conventions the above equation becomes

$$v^{i'} = \frac{\partial x^{i'}}{\partial x^j}\bigg|_{P_o} v^j. \tag{A.7}$$

Equation (A.7) can be written in matrix form as

$$v' = Xv \tag{A.8}$$

where v' is the column matrix

$$\begin{pmatrix} v^{0'} \\ v^{1'} \\ v^{2'} \\ v^{3'} \end{pmatrix}$$

v is the column matrix

$$\begin{pmatrix} v^{0} \\ v^{1} \\ v^{2} \\ v^{3} \end{pmatrix}$$

and X is the 4×4 matrix defined by

$$X = \begin{pmatrix} \frac{\partial x^{0'}}{\partial x^0}|_{P_o} & \frac{\partial x^{0'}}{\partial x^1}|_{P_o} & \frac{\partial x^{0'}}{\partial x^2}|_{P_o} & \frac{\partial x^{0'}}{\partial x^3}|_{P_o} \\ \frac{\partial x^{1'}}{\partial x^0}|_{P_o} & \frac{\partial x^{1'}}{\partial x^1}|_{P_o} & \frac{\partial x^{1'}}{\partial x^2}|_{P_o} & \frac{\partial x^{1'}}{\partial x^3}|_{P_o} \\ \frac{\partial x^{2'}}{\partial x^0}|_{P_o} & \frac{\partial x^{2'}}{\partial x^1}|_{P_o} & \frac{\partial x^{2'}}{\partial x^2}|_{P_o} & \frac{\partial x^{2'}}{\partial x^3}|_{P_o} \\ \frac{\partial x^{3'}}{\partial x^0}|_{P_o} & \frac{\partial x^{3'}}{\partial x^1}|_{P_o} & \frac{\partial x^{3'}}{\partial x^2}|_{P_o} & \frac{\partial x^{3'}}{\partial x^3}|_{P_o} \end{pmatrix}.$$

Given any vector space a second vector space can be constructed, namely the *dual* of the original. Furthermore, each basis for the original vector space induces a dual basis for the dual vector space. The dual of the tangent space $T_{P_o}(M)$ will be denoted by $T^*_{P_o}(M)$. Vectors belonging to $T_{P_o}(M)$ are called *contravariant* vectors and vectors belonging to $T^*_{P_o}(M)$ are called *covariant* vectors. As already introduced above, the components of a contravariant vector, say v, relative to a natural basis are denoted by a Latin 'upstairs' index as v^i. In order to distinguish between components of vectors belonging to $T_{P_o}(M)$ and $T^*_{P_o}(M)$ it is conventional to denote the components of a covariant vector, say w, relative to the basis in $T^*_{P_o}(M)$ induced by the natural basis in $T_{P_o}(M)$ by a Latin 'downstairs' index as w_i. Under a transformation of coordinates these components transform not according to (A.8), but rather according to the equation

$$w' = wX^{-1} \tag{A.9}$$

where w' is the row matrix $(w_0^{'}, w_1^{'}, w_2^{'}, w_3^{'})$, w is the row matrix (w_0, w_1, w_2, w_3) and X^{-1} is the inverse of the matrix X, defined by

$$XX^{-1} = I$$

where I is the unit matrix. The fact that X^{-1} exists follows because X is the Jacobian matrix of the coordinate transformation and is non singular since the transformation is invertible. According to the chain rule

$$\left.\frac{\partial x^{i'}}{\partial x^{i}}\right|_{P_o}\left.\frac{\partial x^{j}}{\partial x^{k'}}\right|_{P_o}=\left.\frac{\partial x^{i'}}{\partial x^{k'}}\right|_{P_o}=\delta^{i'}_{k'}$$

so that the matrix X^{-1} has components

$$\begin{pmatrix} \frac{\partial x^0}{\partial x^{0'}}|_{P_o} & \frac{\partial x^0}{\partial x^{1'}}|_{P_o} & \frac{\partial x^0}{\partial x^{2'}}|_{P_o} & \frac{\partial x^0}{\partial x^{3'}}|_{P_o} \\ \frac{\partial x^1}{\partial x^{0'}}|_{P_o} & \frac{\partial x^1}{\partial x^{1'}}|_{P_o} & \frac{\partial x^1}{\partial x^{2'}}|_{P_o} & \frac{\partial x^1}{\partial x^{3'}}|_{P_o} \\ \frac{\partial x^2}{\partial x^{0'}}|_{P_o} & \frac{\partial x^2}{\partial x^{1'}}|_{P_o} & \frac{\partial x^2}{\partial x^{2'}}|_{P_o} & \frac{\partial x^2}{\partial x^{3'}}|_{P_o} \\ \frac{\partial x^3}{\partial x^{0'}}|_{P_o} & \frac{\partial x^3}{\partial x^{1'}}|_{P_o} & \frac{\partial x^3}{\partial x^{2'}}|_{P_o} & \frac{\partial x^3}{\partial x^{3'}}|_{P_o} \end{pmatrix}.$$

Equation (A.9) can now be written in a form analogous to (A.7), namely

$$w_i^{'}=w_j\left.\frac{\partial x^{j}}{\partial x^{i'}}\right|_{P_o}$$

or, equivalently,

$$w_i^{'}=\left.\frac{\partial x^{j}}{\partial x^{i'}}\right|_{P_o}w_j. \tag{A.10}$$

Remember that the vector spaces $T_{P_o}(M)$ and $T^*_{P_o}(M)$ are defined over $\mathbb{R}$ so that the scalars are just real numbers. Of course the value of a scalar ϕ defined at some point $P_o \in M$ does not depend on the choice of local chart and so under a coordinate transformation a scalar transforms trivially as

$$\phi' = \phi. \tag{A.11}$$

Having defined $T_{P_o}(M)$ and $T^*_{P_o}(M)$, a whole family of vector spaces are constructed by taking the repeated tensor products of $T_{P_o}(M)$ and $T^*_{P_o}(M)$. The tensor product vector space

$$\underbrace{T_{P_o}(M)\times\cdots\times T_{P_o}(M)}_{p\ times}\times\underbrace{T^*_{P_o}(M)\times\cdots\times T^*_{P_o}(M)}_{q\ times}$$

will be denoted by

$$T_{P_o}{}^{(p)}_{(q)}(M),$$

the spaces $T_{P_o}{}^{(0)}_{(0)}(M)$, $T_{P_o}{}^{(1)}_{(0)}(M)$ and $T_{P_o}{}^{(1)}_{(1)}(M)$ coinciding with $\mathbb{R}$, $T_{P_o}(M)$ and $T^*_{P_o}(M)$ respectively. An element of the vector space $T_{P_o}{}^{(p)}_{(q)}(M)$ will be called a tensor of *contravariant valence* p and *covariant valence* q (shortened to a tensor of valence p/q). The natural basis in $T_{P_o}(M)$ induces a basis in $T_{P_o}{}^{(p)}_{(q)}(M)$ (this vector space is of dimension 4^{p+q}) and the components of a tensor, say T, belonging to $T_{P_o}{}^{(p)}_{(q)}(M)$ relative to the basis can be denoted by

$$T^{\overbrace{i\ldots k}^{p\ times}}{}_{\underbrace{l\ldots n}_{q\ times}}.$$

The indices $i, \ldots, k$ are referred to as contravariant indices, the indices $l, \ldots, n$ as covariant indices.In all there are $p+q$ indices and each ranges from 0 to 3. Hence T has 4^{p+q} components as required. Under a transformation of coordinates these components transform according to the equation

$$T'^{i\ldots k}{}_{l\ldots n} = \left.\frac{\partial x^{i'}}{\partial x^p}\right|_{P_o} \cdots \left.\frac{\partial x^{k'}}{\partial x^r}\right|_{P_o} \left.\frac{\partial x^s}{\partial x^{l'}}\right|_{P_o} \cdots \left.\frac{\partial x^u}{\partial x^{n'}}\right|_{P_o} T^{p\ldots r}{}_{s\ldots u}. \tag{A.12}$$

Equations (A.7), (A.10) and (A.11) are all special cases of (A.12). It is sometimes convenient to specify the enumerative indices $i \ldots k$ and $l \ldots n$ collectively and write the components of T as $T^{(p)}{}_{(q)}$. Then equation (A.12) can be written symbolically as

$$T'^{(p)}{}_{(q)} = \left.\frac{\partial x'^{(p)}}{\partial x^{(p)}}\right|_{P_o} \left.\frac{\partial x^{(q)}}{\partial x'^{(q)}}\right|_{P_o} T^{(p)}{}_{(q)}.$$

Each of the vector spaces $T_{P_o}{}^{(p)}_{(q)}(M)$ has amongst its elements a zero vector. These will be referred to, without specifying valence, as the zero tensor $\mathbf{0}$. The components of the zero tensor (of any valence) are, of course, all zero.

A.1.3 *Tensor algebra*

Since tensors of the same valence belong to the same vector space they can be multiplied by scalars (real numbers), added, and, if required, the sum can be equated to the zero tensor. In this way, algebraic tensor equations can be constructed at any point P_o. Such equations have the important property of being defined without reference to a particular coordinate system and so are especially useful in modelling geometrical properties of the space-time manifold and their physical consequences. Nevertheless we shall find it easier to develop most of the theory using components and to write tensor equations as in the following example:

$$6T^{ij}{}_k + P^{ij}{}_k + Q^{ij}{}_k = 0. \tag{A.13}$$

In this equation $T^{ij}{}_k, P^{ij}{}_k$ and $Q^{ij}{}_k$ are the components of three tensors $T, P, Q \in T_{P_o}{}^{(2)}_{(1)}(M)$ respectively. $6T^{ij}{}_k$ will be the components of the tensor $6T$ and $P^{ij}{}_k + Q^{ij}{}_k$ will be the components of the tensor $P+Q$. Of course the values of the components of the tensors depend upon the choice of coordinate system, the important fact about component equations such as (A.13) is that if they hold in *one* coordinate system then they will hold in *all* coordinate systems.

The tensor product of two tensors is a tensor belonging to the appropriate tensor product space and will have components equal to the direct product of the components of the original tensors. For example, consider $R \in T_{P_o}{}^{(1)}_{(0)}(M)$ and $S \in T_{P_o}{}^{(1)}_{(1)}(M)$. Then $U = R \times S \in T_{P_o}{}^{(2)}_{(1)}(M)$ and the tensor U will have components $U^{ij}{}_k$ given by

$$U^{ij}{}_k = R^i S^j{}_k.$$

Such a term $R^i S^j{}_k$ or, similarly, $V^{ij} W_k$ could well have been included in the equation (A.13).

There is one further trick for constructing a 'new' tensor from a given tensor, and this is far less obvious than the sums and products introduced above. As an example of the method consider a tensor $T \in T_{P_o}{}^{(2)}_{(2)}(M)$ having components $T^{ij}{}_{kl}$. Under a change of coordinates these components transform as follows:

$$T^{ij}{}_{kl}{}' = \left.\frac{\partial x^{i'}}{\partial x^s}\right|_{P_o} \left.\frac{\partial x^{j'}}{\partial x^l}\right|_{P_o} \left.\frac{\partial x^u}{\partial x^{k'}}\right|_{P_o} \left.\frac{\partial x^v}{\partial x^{l'}}\right|_{P_o} T^{st}{}_{uv}. \tag{A.14}$$

Now consider the array obtained in each coordinate system by putting one contravariant index equal to a covariant index and summing according to the *Einstein summation convention*, for example, the array defined in the coordinate system x^i by

$$C^i{}_l = T^{ij}{}_{jl}$$

and in the coordinate system $x^{i'}$ by

$$C^i{}_l{}' = T^{ij}{}_{jl}{}'.$$

Then

$$C^i{}_l{}' = \left.\frac{\partial x^{i'}}{\partial x^s}\right|_{P_o} \left.\frac{\partial x^{j'}}{\partial x^t}\right|_{P_o} \left.\frac{\partial x^u}{\partial x^{j'}}\right|_{P_o} \left.\frac{\partial x^v}{\partial x^{l'}}\right|_{P_o} T^{st}{}_{uv}. \tag{A.15}$$

Now, using the chain rule,

$$\left.\frac{\partial x^{j'}}{\partial x^t}\right|_{P_o}\left.\frac{\partial x^u}{\partial x^{j'}}\right|_{P_o} = \left.\frac{\partial x^u}{\partial x^t}\right|_{P_o} = \delta^u_t$$

and

$$\delta^u_t T^{st}{}_{uv} = T^{su}{}_{uv} = C^s{}_v$$

(notice here the Kronecker delta is being used as a substitution operator, substituting the index u for the summed index t). Hence (A.15) becomes

$$C^{i'}{}_{l'} = \left.\frac{\partial x^{i'}}{\partial x^s}\right|_{P_o}\left.\frac{\partial x^v}{\partial x^{l'}}\right|_{P_o} C^s{}_v$$

which is the transformation law for the components of a tensor of contravariant valence one and covariant valence one. It follows that the array $C^i{}_l$ defines a unique tensor $C \in T_{P_o}{}^{(1)}_{(1)}(M)$. It is important to observe that the proof of this result could not have been carried through if, for example, the two contravariant indices had been put equal and summed because then one would have obtained a term

$$\left.\frac{\partial x^{i'}}{\partial x^s}\right|_{P_o}\left.\frac{\partial x^{i'}}{\partial x^t}\right|_{P_o}$$

which *cannot* be simplified using the chain rule. The process of putting one contravariant index equal to a covariant index and summing is called *contraction* and if carried out on a tensor $\in T_{P_o}{}^{(p)}_{(q)}(M)$ will yield a tensor $\in T_{P_o}{}^{(p-1)}_{(q-1)}(M)$. It follows that a term of the form $M^{sij}{}_{sk}$ could well have been included in the equation (A.13).

Two particular examples of contractions are worth mentioning here. If $T \in T_{P_o}{}^{(1)}_{(1)}(M)$ as components $T^i{}_j$ then the contraction $T^i{}_i \in T_{P_o}{}^{(0)}_{(0)}(M)(= \mathbb{R})$, i.e. is a scalar. This scalar is called the *trace* of the tensor T. If $u \in T_{P_o}(M)$ and $v \in T^*_{P_o}(M)$ then the contraction $u^i v_i$ of the tensor $u \times v$ with components $u^i v_j$ is also a scalar (notice that although this looks like a scalar product it does *not* define the scalar product of two vectors belonging to the same vector space).

As already stated, much of the development of the theory given here will be in terms of components and it is very important to note the following three facts concerning the indices which appear in the component formulation of any tensor equation.

(1) a single (free) index appearing in one term of a sum of tensor components *must* appear in the same position (i.e. 'upstairs' or 'downstairs') in every other term of the sum.
(2) Repeated (summation) indices *never* appear on the same level (i.e both 'upstairs' or 'downstairs' — these indices indicate a contraction).
(3) No index can appear more than twice in any term of a sum of tensor components.

If one adds the convention that a derivative index such as i in $\frac{\partial}{\partial x^i}$ is to be considered as a 'downstairs' index, then the above facts are of great help in remembering the positions of the indices in the transformation law for tensor components, equation (A.12).

The ordering of contravariant indices and of covariant indices is very important and indeed it will be seen later that the relative ordering of contravariant and covariant indices is also very important. For example, consider a tensor $T \in T_{P_o}{}^{(2)}_{(0)}(M)$ of contravariant valence two. If the components of the tensor are denoted by T^{ij} then, in general, it will be found that

$$T^{ij} \neq T^{ji}.$$

A tensor T whose components satisfy the equation

$$T^{ij} = T^{ji}$$

is said to be symmetric while a tensor T whose components satisfy the equation

$$T^{ij} = -T^{ji}$$

is said to be skew-symmetric (or antisymmetric). Since

$$T^{ij} = \frac{1}{2}(T^{ij} + T^{ji}) + \frac{1}{2}(T^{ij} - T^{ji})$$

it follows that any tensor $T \in T_{P_o}{}^{(2)}_{(0)}(M)$ can be written as the sum of a symmetric tensor T_+ having components

$$\frac{1}{2}(T^{ij} + T^{ji})$$

and a skew-symmetric tensor T_- having components

$$\frac{1}{2}(T^{ij} - T^{ji})$$

so that

$$T = T_+ + T_-.$$

The following symbols are often used to denote the components of T_+ and T_-:

$$T^{(ij)} = \frac{1}{2}(T^{ij} + T^{ji})$$

and

$$T^{[ij]} = \frac{1}{2}(T^{ij} - T^{ji}).$$

Notice that if $T \in T_{P_o}{}^{(1)}_{(1)}(M)$ and has components $T^i{}_j$ there would be no sense in which one could define T to be symmetric and so write

$$T^i{}_j = T^j{}_i.$$

This equation is contrary to the remarks concerning the positioning of indices and even if the equation were to hold in one coordinate system there is no reason why it should hold true in any other.

Higher valence tensors will often possess symmetry properties which can be expressed as linear relationships between the components of the tensor and arrays obtained by permuting the indices on these components.

A.1.4 *Tensor detection*

Given a particular basis for a vector space V, each vector belonging to V defines a unique set of components relative to the basis. Conversely, each array of numbers (of the appropriate dimension), considered as a set of components, defines a unique vector belonging to V. Hence at any point $P_o \in M$ an array of numbers $T_o{}^{(p)}_{(q)}$ defined in a given coordinate system $x_o{}^i$ defines a unique tensor belonging to $T_{P_o}{}^{(p)}_{(q)}(M)$. The components $T^{(p)}{}_{(q)}$ of this tensor in a general coordinate system x^i are found by applying the appropriate tensor transformation law, thus

$$T^{(p)}{}_{(q)} = \left.\frac{\partial x^{(p)}}{\partial x_o^{(p)}}\right|_{P_o} \left.\frac{\partial x_o^{(q)}}{\partial x^{(q)}}\right|_{P_o} T_o{}^{(p)}{}_{(q)}. \tag{A.16}$$

In the above the array $T_o{}^{(p)}{}_{(q)}$ was defined in a particular coordinate system. If an array is defined in each coordinate system then these arrays will *not* define a unique tensor unless they are related by the appropriate tensor transformation law. Two theorems will now be introduced which will enable one to decide whether these arrays define a unique tensor, without the need to investigate transformation properties. These theorems are called *quotient* theorems. In these theorems all arrays and tensors are defined at a point $P_o \in M$ and all coordinate systems are defined about P_o

Theorem A.1 *If an array of numbers $\hat{S}^{(p)}{}_{(q)}$ is defined in each coordinate system and if the products of $\hat{S}^{(p)}{}_{(q)}$ with the components of a non-zero tensor are themselves the components of a tensor then the arrays $\hat{S}^{(p)}{}_{(q)}$ are the components of a unique tensor ($\in T_{P_o}{}^{(p)}_{(q)}(M)$).*

Proof. By hypothesis

$$\hat{S}^{(p)}{}_{(q)}T^{(r)}{}_{(s)} = R^{(p+r)}{}_{(q+s)} \tag{A.17}$$

where $T^{(r)}{}_{(s)}$ and $R^{(p+r)}{}_{(q+s)}$ are the components of two tensors of the indicated valences. Evaluating this equation in a particular coordinate system $x_o{}^i$ gives

$$\hat{S}_o{}^{(p)}{}_{(q)}T_o{}^{(r)}{}_{(s)} = R_o{}^{(p+r)}{}_{(q+s)}.$$

Transforming back into the general coordinate system using the tensor transformation law yields

$$S^{(p)}{}_{(q)}T^{(r)}{}_{(s)} = R^{(p+r)}{}_{(q+s)} \tag{A.18}$$

where $S^{(p)}{}_{(q)}$ are the components of the unique tensor constructed from the array $\hat{S}_o{}^{(p)}{}_{(q)}$ defined in the coordinate system $x_o{}^i$. Subtracting (A.18) from (A.17) gives

$$[\hat{S}^{(p)}{}_{(q)} - S^{(p)}{}_{(q)}]T^{(r)}{}_{(s)} = 0. \tag{A.19}$$

Since the tensor T is non zero at least one of its components $T^{(r)}{}_{(s)}$ is non zero. Therefore $\hat{S}^{(p)}{}_{(q)} - S^{(p)}{}_{(q)}$ must be zero. Hence

$$\hat{S}^{(p)}{}_{(q)} = S^{(p)}{}_{(q)}$$

and so $\hat{S}^{(p)}{}_{(q)}$ are the components of a unique tensor $\in T_{P_o}{}^{(p)}_{(q)}(M)$. □

Theorem A.2 *If an array of numbers $\hat{S}^{(p)}{}_{(q)}$ is defined in each coordinate system and if the contractions of $\hat{S}^{(p)}{}_{(q)}$ with the components of k arbitrary contravariant and l arbitrary covariant vectors are themselves the components of a tensor then the arrays $\hat{S}^{(p)}{}_{(q)}$ are the components of a unique tensor ($\in T_{P_o}{}^{(p)}_{(q)}(M)$).*

Proof. In order to avoid the complication of having enumerative indices to label the k arbitrary contravariant and l arbitrary covariant vectors, suppose that $k = 2$ and $l = 1$. Then by hypothesis

$$\hat{S}^{i(p-1)}{}_{jk(q-2)}u_i v^j w^k = R^{(p-1)}{}_{(q-2)}$$

where u_i, v^j, w^k are the components of *arbitrary* vectors and $R^{(p-1)}{}_{(q-2)}$ are the components of a tensor of the indicated valence. Following the proof of the last theorem an equation analogous to (A.19) is obtained, namely

$$[\hat{S}^{i(p-1)}{}_{jk(q-2)} - S^{i(p-1)}{}_{jk(q-2)}]u_i v^j w^k = 0. \tag{A.20}$$

Notice the difference between (A.19) and (A.20). In (A.19) the indices on the components $T^{(r)}{}_{(s)}$ are independent of the indices on the terms in the bracket whereas in (A.20) the indices on the components $u_i v^j w^k$ are *contracted* with the indices on the terms in the bracket.

Consider an arbitrary coordinate system and choose as the vectors $\mathbf{u}, \mathbf{v}, \mathbf{w}$ vectors having following components in this coordinate system:

$$u_i = \delta_i^a, v^j = \delta_b^j, w^k = \delta_c^k$$

(i.e. the only non zero component of $\mathbf{u}$ is the component which is equal to one). Then (A.20) becomes

$$[\hat{S}^{i(p-1)}{}_{jk(q-2)} - S^{i(p-1)}{}_{jk(q-2)}]\delta_i^a \delta_b^j \delta_c^k = 0$$

or

$$\hat{S}^{a(p-1)}{}_{bc(q-2)} - S^{a(p-1)}{}_{bc(q-2)} = 0.$$

This is true for all a, b, c and so, in the chosen coordinate system,

$$\hat{S}^{(p)}{}_{(q)} = S^{(p)}{}_{(q)}.$$

Since the coordinate system was chosen arbitrarily this result must hold true in *all* coordinate systems and so $\hat{S}^{(p)}{}_{(q)}$ are the components of a unique tensor belonging to $T_{P_o}{}^{(p)}_{(q)}(M)$. □

A.1.5 *An illustrative example from classical mechanics*

Classical mechanics is based on a three-dimensional Euclidean manifold so that the vectors one meets all have three components. In what follows we shall restrict attention to rectangular Cartesian coordinates. Remember that $\mathbf{r} = x\hat{\mathbf{i}} + y\hat{\mathbf{j}} + z\hat{\mathbf{k}}$ so that, for example, $\frac{\partial \mathbf{r}}{\partial x} = \hat{\mathbf{i}}$. Hence the natural basis vectors are just the usual Cartesian basis vectors $\hat{\mathbf{i}}$, $\hat{\mathbf{j}}$, $\hat{\mathbf{k}}$. Notice that this result is independent of the choice of a point P_o — the tangent spaces $T_{P_o}(M)$ for different points P_o of the manifold are isomorphic, a result which is surely related to the fact that Euclidean vectors have 'no location' in space.

A transformation from one rectangular Cartesian coordinate system to another results in a rotation of the basis vectors $\hat{\mathbf{i}}$, $\hat{\mathbf{j}}$, $\hat{\mathbf{k}}$. It is well known that a rotation is represented by an orthogonal matrix so that the equation corresponding to (A.8) will be

$$v' = Xv \tag{A.21}$$

where v' and v are 3×1 column matrices and X is an orthogonal 3×3 matrix, i.e.

$$X^{-1} = X^t \tag{A.22}$$

where t denotes the transpose. Equation (A.21) is the transformation law for the components of a contravariant vector. The corresponding transformation law for the components of a covariant vector is

$$w' = wX^{-1} \tag{A.23}$$

where w^1 and w are 3×1 row matrices. Taking the transpose of (A.21) gives

$$(v')^t = v^t X^t$$

or, using (A.22)

$$(v')^t = v^t X^{-1}. \tag{A.24}$$

Notice that (A.23) and (A.24) are identical so that the components of a contravariant vector (written as a row matrix rather than as a column matrix) can be considered to be the components of a covariant vector! For each point P_o the two vector spaces $T_{P_o}(M)$ and $T^*_{P_o}(M)$ are isomorphic which is why the concept of contravariance and covariance plays very little role in Euclidean vector algebra. Indeed it is conventional to treat all components as the components of covariant vectors so that vector indices always appear as 'downstairs' indices and the upstairs/downstairs conventions no longer hold.

Two vectors which arise in the study of rigid body mechanics are the angular velocity vector $\mathbf{w}$ and the angular momentum vector $\mathbf{h}$. If the origin is chosen to be fixed in the rigid body then the components of the angular momentum **h** are found to be linear combinations of the components of the angular velocity **w**, i.e.

$$h_i = I_{ij} w_j.$$

Now h_i are the components of a vector for all angular velocity vectors **w** and so by the second quotient theorem the array I_{ij} are the components of a unique tensor. This tensor is called the *moment of inertia tensor* of the rigid body.

A.1.6 *Tensor fields*

The scalars, vectors and tensors introduced above are all defined at some point P_o of the space-time manifold M. In order to develop the theory of general relativity it is necessary to introduce the idea of *fields*. The simplest field is a *scalar field* (not to be confused with a field of scalars). Such a field associates a scalar with each point of the manifold or with each point of some region of the manifold. The scalars are real numbers and so a scalar field is simply a real valued function defined on the manifold (the domain may or may not be the whole of M). If the value of the scalar field ϕ is the same at all points then the scalar field is said to be a constant scalar field. In general the value will vary from point to point and will be denoted by $\phi(x)$ or by $\phi(x^0, \ldots, x^3)$. Notice that $\phi(x^i)$ is not a very good notation because it might denote the four values $\phi(x^0)$, $\phi(x^1)$, $\phi(x^2)$, $\phi(x^3)$!

A tensor field of contravariant valence p and covariant valence q associates a tensor $\in T_{P_o}{}^{(p)}_{(q)}(M)$ with each point P_o of the manifold or with each point P_o of some region of the manifold. Such a tensor field will be denoted by $T(x)$ or $T(x^0, \ldots, x^3)$. The components of a tensor field are real valued functions defined on the manifold and their value at x will be denoted by $T^{(p)}_{(q)}(x)$ or $T^{(p)}_{(q)}(x^0, \ldots, x^3)$. As an example, the four velocity of a moving particle is a contravariant vector field whose domain is the world line of the particle.

The components of a tensor field will, at *each* point P_o, transform under a transformation of coordinates according to the equation (A.12). It is usual therefore when referring to a tensor field to omit the symbols $|_{P_o}$ from this equation. Furthermore, since tensors defined at a single point $P_o \in M$ are never met in general relativity, it is common to use the word tensor to mean a tensor field. We shall follow this common usage occasionally adding the word 'field' in parenthesis.

Consider the scalar field $\phi(x^0, \ldots, x^3)$. Differentiating partially with respect to each coordinate in turn yields four fractions $\frac{\partial \phi}{\partial x^i}$. In a different coordinate system this procedure would have produced four different functions $\frac{\partial \phi'}{\partial x^{i'}}$. Now the value of a scalar field is *independent* of the choice of coordinates so that $\phi' = \phi$ and

$$\frac{\partial \phi'}{\partial x^{i'}} = \frac{\partial \phi}{\partial x^{i'}}.$$

Using the chain rule

$$\frac{\partial \phi'}{\partial x^{i'}} = \frac{\partial x^j}{\partial x^{i'}} \frac{\partial \phi}{\partial x^j}.$$

Comparing this to (A.10) it follows that the partial derivatives $\frac{\partial \phi}{\partial x^i}$ define a unique covariant vector field. This is the generalisation of the usual gradient of a scalar field. In the following, partial derivative will be denoted by a *comma* as follows

$$\phi_{,i} \equiv \frac{\partial \phi}{\partial x^i}.$$

Consider now a covariant vector field with components $v_i(x^0, \ldots, x^3)$. Differentiating each component with respect to each coordinate in turn yields an array of functions:

$$v_{i,j} \equiv \frac{\partial v_i}{\partial x^j}.$$

In a different coordinate system the array

$$v_{i,j}{}' \equiv \frac{\partial v_i{}'}{\partial x^{j'}}$$

is obtained. Now, by the tensor transformation law,

$$v_i{}' = \frac{\partial x^s}{\partial x^{i'}} v_s$$

so that

$$v_{i,j}{}' = \frac{\partial}{\partial x^{j'}} \left(\frac{\partial x^s}{\partial x^{i'}} v_s \right) = \frac{\partial x^s}{\partial x^{i'}} \frac{\partial}{\partial x^{j'}} (v_s) + \frac{\partial^2 x^s}{\partial x^{j'} \partial x^{i'}} v_s.$$

Using the chain rule

$$v_{i,j}{}' = \frac{\partial x^s}{\partial x^{i'}} \frac{\partial x^t}{\partial x^{j'}} \frac{\partial v_s}{\partial x^t} + \frac{\partial^2 x^s}{\partial x^{j'} \partial x^{i'}} v_s$$

or

$$v_{i,j}{}' = \frac{\partial x^s}{\partial x^{i'}} \frac{\partial x^t}{\partial x^{j'}} v_{s,t} + \frac{\partial^2 x^s}{\partial x^{j'} \partial x^{i'}} v_s.$$

The first term on the right-hand side of this equation is the usual tensor transformation law for a tensor (field) of covariant valence two. The appearance of the second term means that the partial derivatives of the components of a vector do *not* define a unique tensor field. Of course, if attention is confined to linear transformations of the coordinates (for example, rotations of rectangular Cartesian coordinates in three-dimensional Euclidean space or the Lorentz transformation in special relativity), then the second derivatives $\partial^2 x^s/\partial x^{j'}\partial x^{i'}$ vanish and a unique tensor field is obtained.

All the familiar laws of physics are expressed as differential equations rather than as algebraic equations. A difficulty now arises because we cannot use the partial derivative to write down tensor differential equations — it is necessary to define a new 'derivative' which is such that the derivative of a tensor is itself a tensor.

A.1.7 *An illustrative example from special relativity*

Consider a freely moving distribution of matter (i.e. no external forces and no pressure) of rest density ρ_o moving with four velocity v^i. Here ρ_o and v^i are a scalar field and a vector field respectively. Using a $3+1$ dimensional notation with $x^1 = x$, $x^2 = y$, $x^3 = z$ and $x^0 = t$, the four velocity decomposes as

$$v^i = \gamma(\mathbf{v}, 1)$$

where γ is the usual Lorentz factor and $\mathbf{v}$ the three-dimensional velocity. Introduce now a contravariant tensor T of valence two defined by

$$T^{ij} = \rho_o v^i v^j. \tag{A.25}$$

Since the Lorentz transformations are linear it follows from the discussion of the last section that the equation

$$T^{ij}{}_{,i} = 0 \tag{A.26}$$

is a tensor equation. Writing out the summation explicitly yields

$$\frac{\partial T^{1j}}{\partial x^1} + \frac{\partial T^{2j}}{\partial x^2} + \frac{\partial T^{3j}}{\partial x^3} + \frac{\partial T^{0j}}{\partial x^0} = 0$$

or

$$\frac{\partial(\rho_o \gamma v_x v^j)}{\partial x} + \frac{\partial(\rho_o \gamma v_y v^j)}{\partial y} + \frac{\partial(\rho_o \gamma v_z v^j)}{\partial z} + \frac{\partial(\rho_o \gamma v^j)}{\partial t} = 0. \tag{A.27}$$

Putting $j = 0$ in this equation and substituting $v^0 = \gamma$ yields

$$\frac{\partial(\rho_o\gamma^2 v_x)}{\partial x} + \frac{\partial(\rho_o\gamma^2 v_y)}{\partial y} + \frac{\partial(\rho_o\gamma^2 v_z)}{\partial z} + \frac{\partial(\rho_o\gamma^2)}{\partial t} = 0.$$

Now $\rho_o\gamma^2$ is just the 'relativistic' density ρ. Hence

$$\frac{\partial(\rho v_x)}{\partial x} + \frac{\partial(\rho v_y)}{\partial y} + \frac{\partial(\rho v_z)}{\partial z} + \frac{\partial(\rho)}{\partial t} = 0 \qquad \text{(A.28)}$$

or using the three-dimensional ∇ operator:

$$\nabla \cdot (\rho\mathbf{v}) + \frac{\partial\rho}{\partial t} = 0. \qquad \text{(A.29)}$$

This is the *equation of continuity* in fluid mechanics. It expresses the conservation of mass or, in the context of special relativity, the conservation of energy (remember $c = 1$).

Putting $j = 1$ in equation (A.27) gives

$$\frac{\partial(\rho_o\gamma^2 v_x v_x)}{\partial x} + \frac{\partial(\rho_o\gamma^2 v_y v_x)}{\partial y} + \frac{\partial(\rho_o\gamma^2 v_z v_x)}{\partial z} + \frac{\partial(\rho_o\gamma^2 v_x)}{\partial t} = 0$$

i.e.

$$v_x\frac{\partial(\rho v_x)}{\partial x} + v_x\frac{\partial(\rho v_y)}{\partial y} + v_x\frac{\partial(\rho v_z)}{\partial z} + \rho v_x\frac{\partial v_x}{\partial x} + \rho v_y\frac{\partial v_x}{\partial y} + \rho v_z\frac{\partial v_x}{\partial z} + v_x\frac{\partial\rho}{\partial t} + \rho\frac{\partial v_x}{\partial t} = 0.$$

Four terms in this equation cancel by virtue of (A.28) leaving

$$p\mathbf{v} \cdot \nabla(v_x) + \rho\frac{\partial v_x}{\partial t} = 0.$$

Putting $j = 2$ and $j = 3$ in equation (A.27) yields two further equations with v_x replaced by v_y and v_z in turn. These three equations can be combined into the single vector equation

$$p\mathbf{v} \cdot \nabla(\mathbf{v}) + \rho\frac{\partial\mathbf{v}}{\partial t} = 0$$

or

$$\rho\frac{d\mathbf{v}}{dt} = 0.$$

This equation is Euler's equation for free flow with zero pressure in fluid dynamics. It expresses the conservation of the linear momentum. Hence the two conservation laws of energy and of linear momentum can be combined into one tensor equation, namely (A.26). The tensor T is called the *energy-momentum tensor*. By analogy to the usual component expression for $\nabla \cdot \mathbf{u}$ the expression $T^{ij}{}_{,i}$ is called the *divergence* of the tensor T. The energy-momentum tensor of the matter distribution therefore has zero divergence.

A.2 Riemannian geometry and tensor analysis

A.2.1 *Introduction*

The existence of the vectors and tensors introduced in Sec. A.1 depend upon the fact that physical space-time is a assumed to be a differentiable manifold M. In special relativity an additional structure is introduced on the manifold, namely a Lorentz scalar product of signature $(1, 3)$. This scalar product defines on M a symmetric tensor field of covariant valence two. Here this tensor field will be denoted by η and relative to an orthonormal basis it has components

$$\eta_{ij} = \begin{pmatrix} +1 & 0 & 0 & 0 \\ 0 & -1 & 0 & 0 \\ 0 & 0 & -1 & 0 \\ 0 & 0 & 0 & -1 \end{pmatrix} \tag{A.30}$$

at *every* point. This tensor field is called the *Minkowski tensor* and is used to define a square distance between two points. Confining attention to neighbouring points the square distance is given by

$$ds^2 = (dx^0)^2 - (dx^1)^2 - (dx^2)^2 - (dx^3)^2 = \eta_{ij} dx^i dx^j. \tag{A.31}$$

Physically the orthonormal basis in which η has components as in (A.30) is the natural basis associated with an inertial system of coordinates; the components of η in any other system of coordinates can be obtained by applying the usual tensor transformation law.

In special relativity it is usual to confine attention to inertial coordinate systems and the misguided might be led to think that general relativity is a generalisation of special relativity to non-inertial coordinate systems. This is not the case, general relativity is a theory of *gravitation*, the one force which is not covered by special relativity.

Assume that a symmetric second-order covariant tensor field g is defined on the space-time manifold M. Given any two vectors $\mathbf{u}, \mathbf{v} \in T_{P_o}(M)$ the tensor field g can be used to define a scalar, namely the scalar

$$g_{ij} u^i v^j, \tag{A.32}$$

the components g_{ij} of the tensor field being evaluated at the point P_o. Furthermore, if it is assumed that the determinant of g_{ij} is non zero (certainly true of η_{ij} in special relativity) then the equation

$$g_{ij}g^{jk} = \delta_i^k \tag{A.33}$$

will have a solution for g^{jk}. In terms of matrices δ_i^k are the components of the unit matrix and g^{jk} will be the components of the matrix inverse of the matrix having components g_{ij}. The inverse of a symmetric matrix is also symmetric so that g^{jk} will be the components of a symmetric second-order contravariant tensor field g^{-1}. Given any two vectors $\mathbf{u}, \mathbf{v} \in T^*_{P_o}(M)$ the expression

$$g^{ij}u_i v_j \tag{A.34}$$

defines a scalar, the components g^{ij} being evaluated at the point P_o.

Given any vector $\mathbf{v} \in T_{P_o}(M)$ the contractions

$$g_{ij}v^j$$

will be the components of a vector belonging to $T^*_{P_o}(M)$. It is common to use the same kernel letter for these components and to write

$$v_j = g_{ij}v^i. \tag{A.35}$$

The vector $\mathbf{v}$ having components v_j is now a vector $\mathbf{v} \in T^*_{P_o}(M)$ and not the original vector $\mathbf{v} \in T_{P_o}(M)$. The process of contracting a contravariant index with the components of g is called 'lowering an index' for obvious reasons and maps $T_{P_o}(M) \to T^*_{P_o}(M)$. The inverse operation of 'raising an index' maps $T^*_{P_o}(M) \to T_{P_o}(M)$ and is carried out using the components of the contravariant tensor field g^{-1}, thus

$$w^j = g^{ij}w_i. \tag{A.36}$$

Raising and lowering indices are indeed inverse operations because the composition of these two operations leads to the original vector, i.e.

$$g^{ij}(g_{ki}v^k) = g^{ij}g_{ki}v^k = \delta^j_k v^k = v^j.$$

Indices on the components of any tensor can be raised and lowered in an analogous fashion:

$$T_i{}^{jk} = g^{js}T_{is}{}^k = g_{pi}T^{pjk}$$

etc.

In a suitable coordinate system the Minkowski tensor η, given by (A.30), is such that $\eta_{ij,k} = 0$ at *every* point. For a general tensor field g the coordinates cannot be chosen so that $g_{ij,k} = 0$ at every point but can be chosen so that $g_{ij,k} = 0$ at any arbitrarily chosen point P_o. This result is crucial to the theory and is proved below.

Theorem A.3 *The coordinate system can be chosen so that, at any given point P_o, $g_{ij,k} = 0$ (such a coordinate system is called a geodesic coordinate system at P_o).*

Proof. First choose a coordinate system x^i such that the coordinates of the point P_o are $x^i = 0$, and suppose $x^{i'}$ is some coordinate system satisfying

$$x^i = x^{i'} - \frac{1}{2}\Gamma^i_{jk}x^{j'}x^{k'}, \tag{A.37}$$

where the coefficients Γ^i_{jk} are constant such that the coordinates of P_o are $x^{i'} = 0$. Under a change of coordinates

$$\begin{aligned} g_{ij,k}{}' &= \frac{\partial g'_{ij}}{\partial x^{k'}} = \frac{\partial}{\partial x^{k'}}\left[\frac{\partial x^s}{\partial x^{i'}}\frac{\partial x^t}{\partial x^{j'}}g_{st}\right] \\ &= \frac{\partial^2 x^s}{\partial x^{k'}\partial x^{i'}}g_{st} + \frac{\partial x^s}{\partial x^{i'}}\frac{\partial^2 x^t}{\partial x^{k'}\partial x^{j'}}g_{st} + \frac{\partial x^s}{\partial x^{i'}}\frac{\partial x^t}{\partial x^{j'}}\frac{\partial x^u}{\partial x^{k'}}g_{st,u}. \end{aligned}$$

From (A.37)

$$\left.\frac{\partial x^i}{\partial x^{p'}}\right|_{P_o} = \delta^i_p$$

and

$$\left.\frac{\partial^2 x^i}{\partial x^{p'}\partial x^{q'}}\right|_{P_o} = -\Gamma^i_{pq}.$$

Hence

$$\begin{aligned} g_{ij,k}{}'|_{P_o} &= -\Gamma^s_{ki}\delta^t g_{st}|_{P_o} - \delta^s_i\Gamma^t_{kj}g_{st}|_{P_o} + \delta^s_i\delta^t_j\delta^u_k g_{st,u}|_{P_o} \\ &= -\Gamma^s_{ki}g_{sj}|_{P_o} - \Gamma^t_{kj}g_{it}|_{P_o} + g_{ij,k}|_{P_o}. \end{aligned}$$

The coordinate system $x^{i'}$ will satisfy the requirements of the theorem provided that the coefficients Γ can be chosen to make the right-hand side of the equation zero. Changing the summation index on the second term to s and interchanging the order of the indices on g in this term yields the following equation for Γ:

$$\Gamma^s_{ki}g_{sj}|_{P_o} + \Gamma^s_{kj}g_{si}|_{P_o} = g_{ij,k}|_{P_o}. \tag{A.38}$$

Cyclically permuting the indices i, j, k on this equation yields

$$\Gamma^s_{ij} g_{sk}|_{P_o} + \Gamma^s_{ik} g_{sj}|_{P_o} = g_{jk,i}|_{P_o} \tag{A.39}$$

and

$$\Gamma^s_{jk} g_{si}|_{P_o} + \Gamma^s_{ji} g_{sk}|_{P_o} = g_{ki,j}|_{P_o}. \tag{A.40}$$

Adding (A.39) to (A.40) and subtracting (A.38) yields (remember that Γ is symmetric on its lower indices)

$$2\Gamma^s_{ij} g_{sk}|_{P_o} = g_{jk,i}|_{P_o} + g_{ki,j}|_{P_o} - g_{ij,k}|_{P_o}.$$

Multiplying by $g^{kt}|_{P_o}$ and remembering that $g_{sk}g^{kt} = \delta^t_s$ yields a unique value for Γ^t_{ij}, namely

$$\Gamma^t_{ij} = \frac{1}{2} g^{kt}|_{P_o}[g_{jk,i}|_{P_o} + g_{ki,j}|_{P_o} - g_{ij,k}|_{P_o}]. \tag{A.41}$$

This completes the proof of the theorem. □

Although Γ found above is unique, the geodesic coordinate systems are not uniquely defined. From the original equation for $g'_{ij,k}$ it can be seen that the vanishing of $g_{ij,k}$ remains invariant under all *linear* transformations so that there exists a whole family of geodesic coordinate systems related to each other by *linear* transformations. Since $|g_{ij}| \neq 0$, a linear transformation exists which will transform the tensor g_{ij} at P_o into a diagonal form with entries ± 1. Einstein was guided to his theory of gravitation by a principle which he called the *principle of equivalence.* According to this principle a uniform gravitational field is mechanically equivalent to an acceleration. Acceleration is of course always defined relative to an observer and the acceleration of a particle can always be made zero by choosing a suitably accelerating observer. The gravitational field in the vicinity of any point $P_o \in (M)$ will be approximately uniform and therefore can be transformed away by a suitable choice of coordinates (i.e. non rotating coordinates associated with a freely falling observer).

In the absence of a gravitational field, we would expect Einstein's theory of gravitation to reduce to his theory of special relativity based on the Minkowski tensor η. It was shown above that geodesic coordinates about an arbitrarily chosen point P_o always exist such that the tensor field g, evaluated at P_o, takes a diagonal form with entries ± 1. If attention is restricted to those tensors g which are such that this diagonal form is

$$diag(+1, -1, -1, -1),$$

then we have a possible mathematical model of the principle of equivalence, the geodesic coordinates in which the tensor g at P_o becomes equal to the Minkowski tensor η corresponding to the freely falling frames of reference in which the gravitational field vanishes. Einstein therefore based his theory of gravitation on a four-dimensional differentiable manifold on which is defined a symmetric covariant second-order tensor field g which reduces, at each point, to the Minkowski tensor η when geodesic coordinates at the point are used. Such a tensor field is said to have a *Lorentzian signature.* The way in which the tensor field g is used to model gravitation will be left to a later section. For the present we shall be interested in investigating the structure of a differentiable manifold with such a tensor field g defined on it. Incidentally, another important principle is the *principle of covariance* which states that physical laws are the same for all observers. Because of this we will be concerned with general coordinate systems and will not pay particular attention to the geodesic coordinate systems.

Given a vector $\mathbf{v} \in T_{P_o}(M)$, the contracted product

$$v^i v^j g_{ij}$$

is a scalar quantity, its value being independent of the choice of coordinate system. In a geodesic coordinate system about P_o the product becomes $v^i v^j \eta_{ij}$ and following the notation used in a Minkowski space the vector $\mathbf{v}$ will be called spacelike, null or timelike, according to as to whether the product is $< 0, 0$ or > 0 respectively. Also, two non zero vectors, $\mathbf{u}, \mathbf{v} \in T_{P_o}(M)$, are said to be orthogonal if

$$u^i v^j g_{ij} = 0.$$

Notice that

$$u^i v^j g_{ij} = u^i (v^j g_{ij}) = u^i v_i$$

or

$$u^i v^j g_{ij} = (u^i g_{ij}) v^j = u_j v^j.$$

In particular,

$$v^i v^j = g_{ij} = v^i v_i.$$

A.2.2 *Covariant derivatives*

In Sec. A.1.6 it was shown that if **v** is a covariant vector field then the partial derivatives

$$v_{i,j}$$

are *not* the components of a unique tensor field unless attention is confined to linear transformations of the coordinates. We saw in the last section that the geodesic coordinate systems $x^{i'}$ at any point $P_o \in (M)$ are related to each other by linear transformations. The partial derivatives

$$v_{i,j}{}'\Big|_{P_o} = \frac{\partial v_i{}'}{\partial x^{j'}}\Big|_{P_o},$$

where $v_i{}'$ are the components of **v** relative to the natural basis vectors associated with the geodesic coordinate system $x^{i'}$, therefore transform as the components of a tensor belonging to $T_{P_o}{}^{(u)}_{(z)}(M)$ under a transformation from one geodesic coordinate system to another. A unique tensor can be constructed (see Sec. A.1.4) by using the tensor transformation law to transform $v_{i,j}{}'|_{P_o}$ to a general coordinate system x^i. This tensor is called the *covariant derivative* of **v** and its components are denoted by a semi-colon (;) as $v_{i;j}|_{P_o}$. By definition

$$\begin{aligned}
v_{i;j}\Big|_{P_o} &= \frac{\partial x^{s'}}{\partial x^i}\Big|_{P_o}\frac{\partial x^{t'}}{\partial x^j}\Big|_{P_o} v_{s,t}{}'\Big|_{P_o} \\
&= \frac{\partial x^{s'}}{\partial x^i}\Big|_{P_o}\frac{\partial x^{t'}}{\partial x^j}\Big|_{P_o}\frac{\partial}{\partial x^{t'}}\left(\frac{\partial x^p}{\partial x^{s'}}v_p\right)\Big|_{P_o} \\
&= \frac{\partial x^{s'}}{\partial x^i}\Big|_{P_o}\frac{\partial x^{t'}}{\partial x^j}\Big|_{P_o}\frac{\partial x^p}{\partial x^{s'}}\Big|_{P_o}\frac{\partial v_p}{\partial x^{t'}}\Big|_{P_o} \\
&\quad + \frac{\partial x^{s'}}{\partial x^i}\Big|_{P_o}\frac{\partial x^{t'}}{\partial x^j}\Big|_{P_o}\frac{\partial^2 x^p}{\partial x^{t'}\partial x^{s'}}\Big|_{P_o} v_p\Big|_{P_o} \\
&= \frac{\partial x^{s'}}{\partial x^i}\Big|_{P_o}\frac{\partial x^{t'}}{\partial x^j}\Big|_{P_o}\frac{\partial x^p}{\partial x^{s'}}\Big|_{P_o}\frac{\partial x^q}{\partial x^{t'}}\Big|_{P_o}\frac{\partial v_p}{\partial x^q}\Big|_{P_o} \\
&\quad + \frac{\partial x^{s'}}{\partial x^i}\Big|_{P_o}\frac{\partial x^{t'}}{\partial x^j}\Big|_{P_o}\frac{\partial^2 x^p}{\partial x^{t'}\partial x^{s'}}\Big|_{P_o} v_p\Big|_{P_o}.
\end{aligned}$$

Using the transformation (A.37) yields

$$v_{i;j}\Big|_{P_o} = \delta_i^s\delta_j^t\delta_s^p\delta_t^q\frac{\partial v_p}{\partial x^q}\Big|_{P_o} - \delta_i^s\delta_j^t\Gamma^p_{ts}v_p\Big|_{P_o},$$

where Γ^p_{ts} is given by equation (A.41). Hence

$$v_{i;j}\Big|_{P_o} = v_{i,j}\Big|_{P_o} - \Gamma^p_{ij} v_p\Big|_{P_o}.$$

(the symmetry of Γ has been used to put the indices i, j in their natural order). This gives the covariant derivative of $\mathbf{v}$ at the point P_o. Dropping P_o gives the covariant derivative of $\mathbf{v}$ as a tensor field, namely

$$v_{i;j} = v_{i,j} - \Gamma^p_{ij} v_p \tag{A.42}$$

where now,

$$\Gamma^p_{ij} = \frac{1}{2} g^{kp}[g_{jk,i} + g_{ik,j} - g_{ij,k}]. \tag{A.43}$$

The quantities Γ^p_{ij} defined by equation (A.43) are called the *Christoffel symbols* for g and are an example of what the differential geometer calls an *affine connexion.*

The covariant derivative of any tensor can now be defined if covariant derivatives are assumed to satisfy the following three axioms:

(1) the usual rule for the derivative of a sum.

(2) the usual rule for the derivative of a product.

(3) the covariant and partial derivatives of a scalar field coincide.

The first two axioms ensure that manipulation of the covariant derivative presents no problems and the third axiom is an obvious axiom to choose when it is remembered that the partial derivative of a scalar field is indeed a covariant vector field.

Consider a contravariant vector field $\mathbf{w}$. For any arbitrary covariant vector field $\mathbf{v}$ the contracted product $v_i w^i$ is a scalar field. Therefore, using axiom 3,

$$(v_i w^i)_{;j} = (v_i w^i)_{,j}.$$

Using axioms 1 and 2,

$$v_{i;j} w^i + v_i w^i{}_{;j} = v_{i,j} w^i + v_i w^i{}_{,j}$$

and, eliminating $v_{i;j}$,

$$v_{i,j} w^i - \Gamma^p_{ij} v_p w^i + v_i w^i{}_{;j} = v_{i,j} w^i + v_i w^i{}_{,j}.$$

Relabelling summation indices so that i always appears as the index on the components of $\mathbf{v}$ yields

$$v_i w^i{}_{;j} = v_i w^i{}_{,j} + \Gamma^i_{pj} v_i w^p.$$

This equation is true for *all* vector fields and therefore

$$w^i{}_{;j} = w^i{}_{,j} + \Gamma^i_{pj} w^p. \tag{A.44}$$

Similarly,

$$T^i{}_{j;k} = T^i{}_{j,k} + \Gamma^i_{pk} T^p{}_j - \Gamma^p_{jk} T^i{}_p \tag{A.45}$$

etc.

Two very useful theorems are now proved.

Theorem A.4 *In a geodesic coordinate system about a point P_o the Christoffel symbols, evaluated at P_o, vanish.*

Proof. The first partial derivatives of the tensor field g vanish at a point P_o when evaluated in a geodesic coordinate system about P_o. The theorem therefore follows trivially from the definition (A.43) of the Christoffel symbols. □

Theorem A.5

$$\delta^i_{j;k} = g_{ij;k} = g^{ij}{}_{;k} = 0.$$

Proof. The components of δ^i_j are constants and so $\delta^i_{j,k} = 0$. In a geodesic coordinate system $g_{ij,k} = 0$ so that differentiating

$$g_{ij} g^{jk} = \delta^k_i$$

in such a coordinate system yields

$$g_{ij} g^{jk}{}_{,l} = 0.$$

Contracting this equation with g^{is} gives

$$\delta^s_j g^{jk}{}_{,l} = g^{sk}{}_{,l} = 0.$$

Hence $\delta^i_{j,k}$, $g_{ij,k}$ and $g^{ij}{}_{,k}$ all vanish in a geodesic coordinate system. Since the partial derivatives coincide with the covariant derivatives in a geodesic coordinate system it follows that

$$\delta^i_{j;k} = g_{ij;k} = g^{ij}{}_{;k} = 0$$

in a geodesic coordinate system. These equations are tensor equations and since they hold in one coordinate system they will hold in every coordinate system. □

Corollary A.1 *Tensor indices can be raised and lowered inside the semi-colon of a covariant derivative. For example,*

$$g^{ij}v_{i;k} = (g^{ij}v_i)_{;k} = v^j{}_{;k}.$$

Note that a covariant derivative index is a tensor index and can be raised in the usual manner, e.g.

$$v_{i;}{}^{k} = g^{kj}v_{i;j}.$$

A.2.3 *The curvature tensor*

A tensor field g and the associated Christoffel symbols defines a Riemannian structure on the space-time manifold. With such a structure the manifold is referred to as a *Riemann space.*

For well behaved scalar fields second-order partial derivatives commute, i.e. if ϕ is such a scalar field then

$$\frac{\partial^2\phi}{\partial x^i \partial x^j} = \frac{\partial^2\phi}{\partial x^j \partial x^i}$$

or

$$\phi_{,ij} = \phi_{,ji}. \tag{A.46}$$

(Note: all indices appearing after a comma are derivative indices). Is the same true of second-order covariant derivatives? Consider

$$\begin{aligned}\phi_{;ij} - \phi_{;ji} &= (\phi_{;i})_{;j} - (\phi_{;ij})_{;i} = (\phi_{,i})_{;j} - (\phi_{,j})_{;i}\\ &= (\phi_{,i})_{,j} - \Gamma^p_{ij}(\phi_{,p}) - (\phi_{,j})_{,i} + \Gamma^p_{ji}(\phi_{,p}).\end{aligned}$$

Now $\phi_{,ij} - \phi_{,ji} = 0$ and $\Gamma^p_{ij} = \Gamma^p_{ji}$ (Γ is symmetric). Hence it follows that

$$\phi_{;ij} - \phi_{;ji} = 0$$

or

$$\phi_{;ij} = \phi_{;ji}. \tag{A.47}$$

Although second covariant derivatives of scalar fields commute, the same is not true in general of second covariant derivatives of tensor fields. For example, consider a covariant vector field $\mathbf{v}$. Then

$$\begin{aligned} v_{i;jk} &= (v_{i;j})_{;k} = (v_{i;j})_{,k} - \Gamma^p_{ik} v_{p;j} - \Gamma^p_{jk} v_{i;p} \\ &= (v_{i,j} - \Gamma^p_{ij} v_p)_{,k} - \Gamma^p_{ik}(v_{p,j} - \Gamma^l_{pj} v_l) - \Gamma^p_{jk} v_{i;p} \\ &= v_{i,jk} - \Gamma^p_{ij,k} v_p - \Gamma^p_{ij} v_{p,k} - \Gamma^p_{ik} v_{p,j} + \Gamma^p_{ik}\Gamma^L_{pj} v_L - \Gamma^p_{jk} v_{i;p}. \end{aligned}$$

It follows that

$$v_{i;jk} - v_{i;kj} = -\Gamma^p_{ij,k} v_p + \Gamma^p_{ik,j} v_p + \Gamma^p_{ik}\Gamma^l_{pj} v_l - \Gamma^p_{ij}\Gamma^l_{pk} v_l.$$

Relabelling summation indices so that the summation index on v is l throughout the right-hand side gives

$$v_{i;jk} - v_{i;kj} = R^l{}_{ijk} v_l \tag{A.48}$$

where

$$R^l{}_{ijk} = \Gamma^l_{ik,j} - \Gamma^l_{ij,k} + \Gamma^p_{ik}\Gamma^l_{pj} - \Gamma^p_{ij}\Gamma^l_{pk}. \tag{A.49}$$

Notice that the coefficients $R^l{}_{ijk}$ are independent of the vector $\mathbf{v}$ and that the left-hand side of (A.48) is a tensor for all vectors $\mathbf{v}$. It follows from the second quotient theorem that the array $R^l{}_{ijk}$ defines a unique tensor (field). This is called the *curvature tensor* or *Riemann tensor* and the identity (A.48) is called the *Ricci identity*.

For a contravariant vector field

$$v^i{}_{;jk} - v^i{}_{;kj} = -R^i{}_{ljk} v^l \tag{A.50}$$

and for a tensor field

$$T^p{}_{q;jk} - T^p{}_{q;kj} = R^l{}_{qjk} T^p{}_l - R^p{}_{ljk} T^l{}_q \tag{A.51}$$

etc.

In the Minkowski space of special relativity coordinates can be chosen so that $\eta_{ij,k} = 0$ everywhere. Then Γ^i_{jk} is zero everywhere so that $R^i{}_{jkl} = 0$. A Riemann space with $R^i{}_{jkl} = 0$ is said to be *flat*.

The symmetry properties of the Riemann tensor are best investigated by choosing an arbitrary point P_o and writing (A.49) in a geodesic coordinate system about O (so that Γ^i_{jk} and $g_{ij,k}$ are both zero at P_o). Then, at P_o,

$$R^l{}_{ijk} = \Gamma^l_{ik,j} - \Gamma^l_{ij,k}$$

or, substituting for the Christoffel symbols

$$R^l{}_{ijk} = \frac{1}{2} g^{ls}(g_{si,kj} + g_{sk,ij} - g_{ik,sj})$$

$$-\frac{1}{2}g^{ls}(g_{si,jk} + g_{sj,ik} - g_{ij,sk}).$$

Lowering the index l by contracting with g_{lh} and remembering that partial derivatives commute gives

$$R_{hijk} = \frac{1}{2}[g_{hk,ij} - g_{ik,hj} - g_{hj,ik} + g_{ij,hk}]. \tag{A.52}$$

By inspection it is seen that the Riemann tensor has the following symmetries: skew on the second two indices:

$$R_{hijk} = -R_{ihjk}, \tag{A.53}$$

skew on the last two indices:

$$R_{hijk} = -R_{hikj}, \tag{A.54}$$

symmetric on interchange of pairs of indices:

$$R_{hijk} = R_{jkhi}, \tag{A.55}$$

cyclic symmetry:

$$R_{hijk} + R_{hjki} + R_{hkij} = 0. \tag{A.56}$$

Of course the equations (A.53)–(A.56) have only been proved in a geodesic coordinate system about a point P_o. However, since the equations are all tensor equations they must also hold in any other coordinate system and since P_o was chosen arbitrarily they must hold at all points. A general fourth covariant order tensor in M would have $4 \times 4 \times 4 \times 4 = 256$ independent components. Because of the symmetries the curvature tensor R will have far fewer independent components.

Theorem A.6 *In the four-dimensional Riemann space of general relativity, the Riemann tensor has* 20 *independent components.*

Proof. (i) consider the components with h, i, j, k all different. There is only one choice for these indices and only two arrangements leading to two independent components.

(ii) consider the components with h and j identical but i and k different. There are four choices for h and $\frac{3\times 2}{2}$ choices for i and k. For each choice there is only one arrangement leading to 12 independent components.

(iii) consider the components with h and j identical and i and k identical. There are $\frac{4\times 3}{2}$ choices for these two pairs. For each choice there is only one arrangement leading to six independent components.

Adding the above yields 20 independent components. □

The valence two tensor field R_{ij} obtained by contraction, namely

$$R_{ij} = R^k{}_{ijk} = -R^k{}_{ikj} \tag{A.57}$$

is called the *Ricci tensor.* Contracting the identity

$$R_{hijk} = R_{kjih}$$

on h and k (i.e. after raising h) yields

$$R_{ij} = R_{ji}$$

so that the Ricci tensor is a symmetric tensor. The scalar defined by

$$R = R^i{}_i \tag{A.58}$$

is called the *curvature scalar* or *Ricci scalar.*

A.2.4 *The Bianchi identities*

Following the method used in deriving the symmetries of the Riemann tensor we evaluate $R^l_{ijk;m}$ at a point P_o in a geodesic coordinate system about P_o. Then

$$R^l{}_{ijk;m} = R^l{}_{ijk,m} = \Gamma^l{}_{ik,jm} - \Gamma^l{}_{ij,km}.$$

Similarly,

$$R^l{}_{ikm;j} = \Gamma^l{}_{im,kj} - \Gamma^l{}_{ik,mj}$$

and

$$R^l{}_{imj;k} = \Gamma^l{}_{ij,mk} - \Gamma^l{}_{im,jk}.$$

Adding these last equations and commuting partial derivatives yields

$$R^l{}_{ijk;m} + R^l{}_{ikm;j} + R^l{}_{imj;k} = 0. \tag{A.59}$$

Again this is a tensor equation and, since it holds in a geodesic coordinate system it will hold in all coordinate systems. Furthermore P_o was chosen arbitrarily and so the equation will hold at all points.

The relations (A.59) are called the *Bianchi identities.* Contracting on l, m yields

$$R^l{}_{ijk;l} + R_{ik;j} - R_{ij;k} = 0.$$

On raising i this equation becomes

$$R^{li}{}_{jk;l} + R^i{}_{k;j} - R^i{}_{j;k} = 0.$$

Contracting on i, j gives

$$R^l{}_{k;l} + R^i{}_{k;i} - R_{;k} = 0,$$

an equation which can be written as

$$\left[R^i{}_k - \frac{1}{2}\delta^i_k R\right]_{;i} = 0. \tag{A.60}$$

The relations (A.60) are called the *contracted Bianchi identities* and play a crucial role in the theory of general relativity. The tensor G is defined by

$$G^i{}_k = R^i{}_k - \frac{1}{2}\delta^i_k R \tag{A.61}$$

is called the *Einstein tensor* and, according to (A.60), has zero divergence (see Sec. A.1.7). Notice that lowering the index on the Kronecker delta yields

$$g_{ij}\delta^i_k = g_{kj},$$

i.e. yields the metric tensor g. Hence the covariant form of the Einstein tensor is

$$G_{ik} = R_{ik} - \frac{1}{2}g_{ik}R. \tag{A.62}$$

A.2.5 *Riemannian geometry and geodesics*

According to the chain rule the coordinate differentials transform, under a change of coordinates, as

$$dx^{i'} = \frac{\partial x^{i'}}{\partial x^j}dx^j,$$

i.e. they form a contravariant vector field. It follows that the contraction $g_{ij}dx^i dx^j$ is a scalar and can be used to define a 'square distance' ds^2 between infinitesimally close points, i.e.

$$ds^2 = g_{ij}dx^i dx^j. \tag{A.63}$$

In this context the tensor g is called the *metric tensor*. The geometry of a differentiable manifold with such a distance defined is called *Riemannian geometry*.

Consider a curve described by a parameter λ. The contravariant vector field

$$v^i = \frac{dx^i}{d\lambda}$$

is tangent to the curve at each point. In Euclidean geometry a curve C is called a straight line if and only if its tangent vector is always fixed in direction, i.e. if and only if the rate of change of the tangent vector along the curve is parallel to the tangent vector itself. Hence the equation of a straight line takes the form

$$\frac{dv^i}{d\lambda} = \alpha v^i.$$

Now

$$\frac{dv^i}{d\lambda} = \frac{\partial v^i}{\partial x^j}\frac{dx^j}{d\lambda} = \frac{\partial v^i}{\partial x^j}v^j = v^i{}_{,j}v^j$$

so that the equation of a straight line can be written as

$$v^i{}_{,j}v^j = \alpha v^i.$$

The tensor equation

$$v^i{}_{;j}v^j = \alpha v^i \tag{A.64}$$

defines a curve in Riemannian geometry which is analogous to the straight line in Euclidean geometry. Such a curve is called a *geodesic* of the Riemann space.

If the parameter describing the curve C is changed from λ to λ' then the tangent vector will change from v^i to $v^{i'}$ where

$$v^i = \frac{dx^i}{d\lambda} = \frac{dx^i}{d\lambda'}\frac{d\lambda'}{d\lambda} = v^{i'}\frac{d\lambda'}{d\lambda}$$

(this is *not* a coordinate transformation). Substituting this into the geodesic equation yields

$$\left(v^{i'}\frac{d\lambda'}{d\lambda}\right)_{;j}v^j = \alpha v^{i'}\frac{d\lambda'}{d\lambda}$$

or

$$v^{i'}{}_{;j}v^{j'}\left(\frac{d\lambda'}{d\lambda}\right)^2 + v^{i'}\frac{d^2\lambda'}{d\lambda^2} = \alpha v^{i'}\frac{d\lambda'}{d\lambda}.$$

Hence if the new parameter is chosen to satisfy the differential equation

$$\frac{d^2\lambda'}{d\lambda^2} = \alpha \frac{d\lambda'}{d\lambda} \tag{A.65}$$

then

$$v^{i'}{}_{;j'}v^{j'} = 0.$$

In other words, the parameter along a geodesic can always be chosen such that α in (A.64) is identically zero. Such a parameter is called a *preferred parameter* or *affine parameter*. The general solution to (A.65) is

$$\lambda' = a + b \int e^{\int \alpha d\lambda} d\lambda$$

so that the affine parameter is defined up to two constants of integration. The constant a corresponds to a change in origin and the constant b to a change in scale.

Of course a straight line in Euclidean geometry is a curve of minimal length and it would, at first sight, have been more natural in developing the concept of a geodesic to have defined a geodesic to have an analogous property. There are some difficulties. The square distance between two points on a general curve c corresponding to λ and $\lambda + d\lambda$ will be given by

$$ds^2 = g_{ij}dx^i dx^j = g_{ij}\frac{dx^i}{d\lambda}d\lambda\frac{dx^j}{d\lambda}d\lambda = g_{ij}v^i v^j d\lambda^2.$$

The distance D along the curve C from a point A to a point B will have to be found by integration. Unfortunately, only if $\mathbf{v}$ is timelike (the curve C is then said to be timelike) can we take the square root of the above to write

$$D = \int_{\lambda_a}^{\lambda_b} (g_{ij}v^i v^j)^{\frac{1}{2}} d\lambda.$$

Indeed, if $\mathbf{v}$ is null then the distance along C is always zero (the curve C is then said to be a *null curve*). Despite these difficulties the geodesics can be defined by a variational principle because if λ is an affine parameter for a geodesic passing through the points A and B then the integral

$$\int_{\lambda_a}^{\lambda_b} g_{ij}v^i v^j d\lambda$$

is stationary for variations of the curve leaving the end points fixed. To prove this result apply the Euler–Lagrange equation in the form

$$\frac{d}{d\lambda}\left(\frac{dL}{d\dot{x}^k}\right) - \frac{dL}{dx^k} = 0,$$

where $L = g_{ij}\dot{x}^i\dot{x}^j$. Then

$$\frac{d}{d\lambda}(2g_{ik}\dot{x}^i) - g_{ij,k}\dot{x}^i\dot{x}^j = 0$$

or

$$2g_{ik}\ddot{x}^i + 2g_{ik,j}\dot{x}^j\dot{x}^i - g_{ij,k}\dot{x}^i\dot{x}^j = 0.$$

Writing $2g_{ik,j}\dot{x}^j\dot{x}^i$ as $(g_{ik,j} + g_{jk,i})\dot{x}^i\dot{x}^j$ and contracting by $\frac{1}{2}g^{ks}$ yields

$$\ddot{x}^s + \frac{1}{2}g^{ks}(g_{ik,j} + g_{jk,i} - g_{ij,k})\dot{x}^i\dot{x}^j = 0$$

or

$$\ddot{x}^s + \Gamma^s_{ij}\dot{x}^i\dot{x}^j = 0. \tag{A.66}$$

This is just the equation (A.64) written out explicitly with $v^i = \dot{x}^i$ and $\alpha = 0$.

A.3 General relativity

A.3.1 *The field equations*

Consider a particle moving freely past a massive object. In the Newtonian theory of gravitation the massive object is replaced by a force ('gravity') and then, since the particle is no longer moving freely, one finds that its path is no longer a straight line. In Einstein's theory the particle is still considered to be moving freely but in a space-time manifold which is, in some way, determined by the massive object. The space-time manifold is assumed to be a Riemann space with metric tensor g of Lorentzian signature and in such a space it is natural to presuppose that the path of a freely moving particle is modelled by a geodesic, such a curve being the analogue of a straight line.

If there is no gravitational field present, and if an inertial coordinate system is being used, the geodesic path of a freely moving particle should reduce to a straight line. This is only possible if a coordinate system can be found in which the Christoffel symbols and therefore the partial derivative of the metric tensor vanish at *all* points of the manifold. The Riemann space then must be a Minkowski space. Even in a Minkowski space the Christoffel symbols will not, in a general coordinate system, vanish. In a general coordinate system the path of the freely moving particle will be given by

$$\frac{d^2x^i}{d\lambda^2} + \Gamma^i_{jk}\frac{dx^j}{d\lambda}\frac{dx^k}{d\lambda} = 0, \tag{A.67}$$

the presence of the term involving the Christoffel symbol indicating the existence of an inertial force and the use of non-inertial coordinates.

In a general Riemann space the geodesics will *not* reduce to straight lines, no matter how the coordinates are chosen. This indicates the presence of a gravitational field. The Christoffel symbols vanish at any chosen point when evaluated in a geodesic coordinate system at the point. This is the mathematical expression of the principle of equivalence.

In the presence of a gravitational field the metric tensor g must, in some way, be determined by the matter distribution which is the source of the gravitational field. In the Newtonian theory the gravitational field is characterised by a scalar potential ϕ satisfying Poisson's equation

$$\nabla^2\phi = -4\pi\gamma\rho, \tag{A.68}$$

where ρ is the density of the matter distribution and γ is a constant. In the special theory of relativity ρ is just part of the energy-momentum tensor T of the distribution and so one might expect T^{ij} to appear on the right-hand side of the field equations of general relativity. The metric tensor components g_{ij} are the analogues of the scalar potential and so again one might expect the left-hand side of the field equations to involve a valence two tensor constructed from the components g_{ij} and their first two derivatives (i.e. since ∇^2 is a second derivative). Einstein was guided to his choice of this valence two tensor by the fact that, in the absence of external forces, the energy-momentum tensor has zero divergence, a condition which can be written in tensor form as (see Sec. A.1.7)

$$T^{ij}{}_{;i} = 0.$$

It was shown in Sec. A.2.4 that the Einstein tensor

$$G^{ij} = R^{ij} - \frac{1}{2}g^{ij}R,$$

which is linear in the second derivatives of the metric tensor, also has zero divergence so that Einstein postulated the field equation

$$G^{ij} = kT^{ij}, \tag{A.69}$$

were k is a constant. In any region of space unoccupied by matter the tensor T^{ij} becomes zero so that (A.69) gives

$$G^{ij} = R^{ij} - \frac{1}{2}g^{ij}R = 0.$$

Contracting on i and j yields $R - \frac{1}{2}4R = 0$ so that $R = 0$. It follows that the Ricci tensor itself must be zero

$$R^{ij} = 0. \tag{A.70}$$

The equation (A.70) is called the *empty space* or *vacuum field equation.*

The Einstein tensor G^{ij} is not the only valence two tensor having zero divergence. A trivial generalisation of this tensor is obtained by adding a term λg^{ij}, where λ is a constant (remember that $g^{ij}{}_{;k} = 0$ so that $(\lambda g^{ij})_{;i} = 0$. This modified tensor is sometimes used in the field equations (A.69), the term λg^{ij} being called the *cosmological term* for reasons which we cannot go into now. In fact it is generally accepted that the cosmological term should *not* be included in the field equations. The field equations give ten equations in the ten unknown components of g_{ij}. If two metrics differ by a coordinate transformation they must represent the same physical situation. For this reason one would expect the solution to (A.69) to involve *four* arbitrary functions, corresponding to the arbitrariness in a general coordinate transformation. This situation would be impossible if the field equations were independent. However, this is not so since the identical vanishing of the covariant divergence of both sides of (A.69) yields four relations between the equations.

From the field equations it can actually be proved that the path of a free particle is a geodesic. However, the geodesic nature of the path must be postulated:

Postulate 1. The path of a freely moving particle (or of a test particle) is a timelike geodesic:

$$\frac{d^2x^i}{ds^2} + \Gamma^i_{jk}\frac{dx^j}{ds}\frac{dx^k}{ds} = 0, \quad g_{jk}\frac{dx^j}{ds}\frac{dx^k}{ds} > 0$$

(note that for a timelike geodesic s can be chosen as an affine parameter).

Postulate 2. The path of a photon (i.e. a light ray) is a null geodesic:

$$\frac{d^2x^i}{d\lambda^2} + \Gamma^i_{jk}\frac{dx^j}{d\lambda}\frac{dx^k}{d\lambda} = 0, \quad g_{jk}\frac{dx^j}{d\lambda}\frac{dx^k}{d\lambda} = 0.$$

(note λ is an affine parameter).

The general theory of relativity is based on the field equation (A.69) together with these two postulates.

A.3.2 *The Newtonian approximation*

Newton's theory of gravitation agrees very well with experiment and it is therefore important that Einstein's theory should approximate to Newton's theory at least for weak static gravitational fields.

The space-time manifold of a *weak* gravitational field will, approximately, be a Minkowski space. Hence coordinates $x^1 = x,\ x^2 = y,\ x^3 = z,\ x^0 = ct$ can be chosen such that

$$g_{ij} \simeq diag(1, -1, -1, -1).$$

If Greek indices $\alpha, \beta, \ldots$ are assumed to range from 0 to 3 this approximation can be written as

$$\begin{aligned} g_{\alpha\beta} &\simeq -\delta_{\alpha\beta},\ g_{\alpha 0} \simeq 0,\ g_{00} \simeq 1. \\ g^{\alpha\beta} &\simeq -\delta_{\alpha\beta},\ g^{\alpha 0} \simeq 0,\ g^{00} \simeq 1. \end{aligned} \tag{A.71}$$

For a *static* gravitational field

$$g_{ij,0} = 0 \tag{A.72}$$

and since the gravitational field is weak in some finite region of the space-time manifold all other partial derivatives of g_{ij} must be approximately zero,

$$g_{ij,0} \simeq 0. \tag{A.73}$$

Now consider a freely moving particle, moving slowly in the above gravitational field. Along the path of the particle

$$ds^2 \simeq -dx^2 - dy^2 - dz^2 + c^2 dt^2.$$

Now

$$dx^2 + dy^2 + dz^2 = \left[\left(\frac{dx}{dt}\right)^2 + \left(\frac{dy}{dt}\right)^2 + \left(\frac{dz}{dt}\right)^2\right] dt^2 = v^2 dt^2.$$

For a slow moving particle $v^2 << c^2$ and so

$$ds^2 \simeq c^2 dt^2. \tag{A.74}$$

The four velocity of the particle will be given by

$$v^\alpha = \frac{dx^\alpha}{ds} = \frac{dx^\alpha}{dt}\frac{dt}{ds} = \frac{dx^\alpha}{dt}\frac{1}{c} \simeq 0$$
$$v^0 = \frac{dx^0}{ds} = \frac{d(ct)}{ds} \simeq 1. \tag{A.75}$$

According to Postulate 1 the path of the particle is a timelike geodesic with equation

$$\frac{d^2x^i}{ds^2} + \Gamma^i_{jk}\frac{dx^j}{ds}\frac{dx^0}{ds} = 0. \tag{A.76}$$

The first three components of (A.76) approximate, using (A.74) and (A.75), to the equation

$$\frac{d^2x^\alpha}{dt^2} + c^2\Gamma^\alpha_{00} = 0. \tag{A.77}$$

Calculating Γ^α_{00} gives

$$\begin{aligned}\Gamma^\alpha_{00} &= \frac{1}{2}g^{\alpha i}[g_{i0,0} + g_{i0,0} - g_{00,i}] \\ &= -\frac{1}{2}g^{\alpha i}g_{00,i} \qquad \text{using(A.72)} \\ &\simeq -\frac{1}{2}g^{\alpha\beta}g_{00,\beta} \simeq +\frac{1}{2}\delta_{\alpha\beta}g_{00,\beta} \qquad \text{using (A.71)} \\ &\simeq \frac{1}{2}g_{00,\alpha}.\end{aligned}$$

Equation (A.77) therefore becomes

$$\frac{d^2x^\alpha}{dt^2} \simeq -\frac{1}{2}c^2 g_{00,\alpha}. \tag{A.78}$$

The Newtonian equation of motion of a particle moving under a gravitational potential ϕ alone is

$$\ddot{\mathbf{r}} = -\nabla\phi$$

so that the Newtonian equation of motion is obtained from the approximation (A.78) provided that

$$\frac{1}{2}c^2 g_{00} = \phi + \text{ constant}.$$

Since $g_{00} = 1$ when no gravitational field is present the constant must be equal to $\frac{1}{2}c^2$. Hence

$$g_{00} = 1 + \frac{2\phi}{c^2}. \tag{A.79}$$

In Newtonian theory the gravitational potential ϕ satisfies Poisson's equation

$$\nabla^2\phi = -4\pi\rho\gamma \tag{A.80}$$

where ρ is the density of the matter distribution (the source of the gravitational field) and γ is the universal gravitational constant. To see that Poisson's equation can be deduced as an approximation consider the Einstein field equations in the form

$$R^{ij} - \frac{1}{2}g^{ij}R = kT^{ij} = k\rho c^2 v^i v^j \tag{A.81}$$

where ρ is the rest density of the matter distribution and v^i is the four velocity of the particles of the distribution. Contracting (A.81) on i and j gives

$$R^i{}_i - \frac{1}{2}\delta^i_i R = k\rho c^2 v^i v_i$$

or

$$R - 2R = k\rho c^2,$$

since $v^i v_i = 1$. Hence

$$R = -k\rho c^2.$$

Substituting for R into (A.81) yields an alternative form for the field equations, namely

$$R^{ij} = k\rho c^2\Big(v^i v^j - \frac{1}{2}g^{ij}\Big).$$

The 00 component of this equation is

$$R^{00} = k\rho c^2(1 - \frac{1}{2}) = \frac{1}{2}k\rho c^2. \tag{A.82}$$

It is convenient to use the fact that

$$R^{00} = g^{0i}g^{0j}R_{ij} \simeq R^{00}$$

to rewrite (A.82) as

$$R_{00} = \frac{1}{2}k\rho c^2. \tag{A.83}$$

Using the definition of the Ricci tensor in terms of the Christoffel symbols yields

$$R_{00} = \Gamma^i_{0i,0} - \Gamma^i_{00,i} + \Gamma^i_{0j}\Gamma^i_{0i} - \Gamma^i_{00}\Gamma^j_{ij}.$$

The first term is zero since the gravitational field is static and the last two terms, which involve squares of the first derivatives of g, will be small compared to $\Gamma^i_{00,i}$. Hence

$$R_{00} \simeq -\Gamma^i_{00,i}$$

and so, remembering that $\Gamma^0_{00,0} = 0$, equation (A.83) becomes approximately equal to

$$-\Gamma^\alpha_{00,\alpha} = \frac{1}{2}k\rho c^2$$

or, using the previous expression for Γ^α_{00},

$$-\frac{1}{2}g_{00,\alpha\alpha} = \frac{1}{2}k\rho c^2.$$

Since g_{00} is related to the gravitational potential ϕ by the equation (A.79), it follows that

$$\frac{1}{c^2}\phi_{,\alpha\alpha} = -\frac{1}{2}k\rho c^2$$

or

$$\nabla^2\phi = -\frac{1}{2}k\rho c^4. \tag{A.84}$$

Comparing (A.84) with (A.80) it is seen that the Newtonian field equations are obtained as an approximation provided that

$$-\frac{1}{2}k\rho c^4 = -4\pi\rho\gamma,$$

or

$$k = 8\pi\gamma/c^4. \tag{A.85}$$

Thus Newton's law of gravitation is included as an approximation to Einstein's theory and equation (A.85) relates the content of proportionality k which appears in Einstein's field equations to the gravitational constant γ.

A.3.3 The Schwarzschild solution

In this section we will consider the gravitational field of a single static gravitating particle. The space-time manifold modelling this gravitational field will be spherically symmetric and so best described in terms of spherical polar coordinates (r, θ, ϕ). In terms of such coordinates the usual Minkowski square distance

$$ds^2 = -dx^2 - dy^2 - dz^2 + c^2 dt^2$$

is given by

$$ds^2 = -dr^2 - r^2(d\theta^2 + \sin^2\theta d\phi^2) + c^2 dt^2.$$

This expression can be generalised, keeping the spherical symmetry (i.e. independence on θ and ϕ), to

$$ds^2 = -p(r)dr^2 - q(r)(d\theta^2 + \sin^2\theta d\phi^2) + c^2 s(r)dt^2.$$

Part of this generalisation is only an apparent generalisation because the function $q(r)$ can be reduced to the original form r^2 simply by defining a new radial coordinate $r' = (q(r))^{\frac{1}{2}}$. Then

$$ds^2 = -p(r)dr^2 - r^2(d\theta^2 + \sin^2\theta d\phi^2) + c^2 s(r)dt^2.$$

Of course, in order to obtain the correct signature, the functions $p(r)$ and $s(r)$ must be positive and this is ensured by writing

$$p(r) = e^{\lambda(r)}$$

and

$$s(r) = e^{\nu(r)}.$$

Then

$$ds^2 = -e^{\lambda(r)}dr^2 - r^2(d\theta^2 + \sin^2\theta d\phi^2) + c^2 e^{\nu(r)}dt^2. \tag{A.86}$$

The expression (A.86) gives the square distance for the most general spherically symmetric, static space-time manifold, a result which we will not actually prove here.

Outside the central particle the field equation is the empty space field equation (A.70), namely

$$R^{ij} = 0$$

or, more usefully,

$$R^i{}_j = 0. \tag{A.87}$$

By direct calculation one obtains

$$R^i{}_j = 0 \quad i \neq j$$

$$R^1{}_1 = -e^{-\lambda}\Big(\frac{\nu'}{r} + \frac{1}{r^2}\Big) + \frac{1}{r^2} = 0, \tag{A.88}$$

$$R^2{}_2 = R^3{}_3 = -e^{-\lambda}\Big(\frac{\nu''}{2} + \frac{\nu'^2}{4} - \frac{\nu'\lambda'}{4} + \frac{\lambda'}{r}\Big) = 0, \tag{A.89}$$

$$R^0{}_0 = e^{-\lambda}\Big(\frac{\lambda'}{r} - \frac{1}{r^2}\Big) + \frac{1}{r^2} = 0, \tag{A.90}$$

where $'$ denotes differentiation with respect to r. Equation (A.90) can be rewritten as

$$\Big(re^{-\lambda}\Big)' = 1$$

so that

$$e^{-\lambda} = 1 - a/r \tag{A.91}$$

where a is a constant of integration. Substituting this into (A.88) gives

$$\nu' = \frac{a}{(r-a)r}$$

so that

$$e^{\nu} = b(1 - a/r). \tag{A.92}$$

The final field equation, (A.89), is identically satisfied by virtue of (A.91) and (A.92). Substituting (A.91) and (A.92) into (A.86) yields

$$ds^2 = -\frac{dr^2}{1 - a/r} - r^2(d\theta^2 + \sin^2\theta d\phi^2) + c^2 b(1 - a/r)dt^2.$$

As r tends to infinity the gravitational effect of the particle will vanish and the space-time geometry should become Minkowskian. Hence $b = 1$ and then

$$ds^2 = -\frac{dr^2}{1 - a/r} - r^2(d\theta^2 + \sin^2\theta d\phi^2) + c^2(1 - a/r)dt^2. \tag{A.93}$$

For weak fields (i.e. for large r) we saw in the last section that g_{00} is connected to the Newtonian potential ϕ by the equation

$$g_{00} = \frac{2\phi}{c^2} + 1.$$

For a single gravitating particle of mass m, the potential ϕ is given by

$$\phi = -\gamma m/r$$

so that

$$g_{00} = 1 - 2\gamma m/c^2 r.$$

Comparing this with (A.93) gives

$$a = 2\gamma m/c^2. \tag{A.94}$$

The space-time manifold with squared distance (A.93) is called the *Schwarzschild solution.* We shall conclude by deriving three general relativistic results which differ markedly from their Newtonian analogues.

A.3.4 *Planetary orbits*

Consider a planet orbiting around the sun. If the effect of the planet on the gravitational field is neglected then the field will be modelled by the Schwarzschild solution (A.93). The orbit of the planet will be a time-like geodesic, the equations of which can be found by applying the Euler–Lagrange equations to the integral

$$\int \left[-\frac{\dot{r}^2}{1-a/r} - r^2\left(\dot{\theta}^2 + \sin^2\theta\dot{\phi}^2\right) + c^2(1-a/r)\dot{t}^2\right] ds$$

where the dot denotes differentiation with respect to s. Since t and ϕ are ignorable one obtains

$$(1-a/r)\dot{t} = \text{constant.} \tag{A.95}$$

and

$$r^2\sin^2\theta\dot{\phi} = \text{constant.} \tag{A.96}$$

The θ equation is

$$\frac{d}{ds}\left(-2r^2\dot{\theta}\right) + 2r^2\sin\theta\cos\theta\dot{\phi}^2 = 0 \tag{A.97}$$

and finally, since the tangent vector to the orbit with parameters is a *unit* timelike vector,

$$\frac{-\dot{r}^2}{1-a/r} - r^2(\dot{\theta} + \sin^2\theta\dot{\phi}^2) + c^2(1-a/r)t^2 = 1, \tag{A.98}$$

an equation analogous to the conservation of energy in Lagrangian mechanics.

In Sec. A.3.3 nothing was said about choosing the axis of the spherical polar coordinates so that we are free to choose it so that initially the planet is moving in the plane $\theta = \frac{\pi}{2}$. Then $\dot{\theta} = \cos\theta = 0$ and substituting this into (A.97) gives $\ddot{\theta} = 0$ initially. Hence the planet *remains* in the plane $\theta = \frac{\pi}{2}$, i.e. the orbit lies in the plane, just as in the Newtonian theory. Substituting $\theta = \frac{\pi}{2}$ into (A.95), (A.96) and (A.98) yields

$$\dot{t} = k(1-a/r)^{-1}/c \tag{A.99}$$

$$r^2\dot{\phi} = h/c \tag{A.100}$$

$$-\frac{\dot{r}^2}{1-a/r} - r^2\dot{\phi}^2 + c^2(1-a/r)\dot{t}^2 = 1. \tag{A.101}$$

The speed of light has been included in the constants in (A.99) and (A.100) in order that h approximates to the Newtonian angular momentum and k to unity (remember that r and ϕ are now plane polar coordinates in the plane of the orbit). Eliminating $\dot{\phi}$ and $\dot{t}$ from (A.101) yields

$$\frac{-\dot{r}^2}{1-a/r} - \frac{h^2}{r^2c^2} + \frac{k^2}{1-a/r} = 1.$$

Following the usual Newtonian treatment we will put $u = \frac{1}{r}$ and

$$\frac{d}{ds} = \frac{d\phi}{dt}\frac{d}{d\phi} = \frac{h}{cr^2}\frac{d}{d\phi}$$

so that

$$-\frac{h^2}{c^2(1-au)}\left(\frac{du}{d\phi}^2\right) - \frac{h^2u^2}{c^2} + \frac{k^2}{1-au} = 1$$

or

$$\left(\frac{du}{d\phi}\right)^2 = -u^2(1-au) + \frac{k^2c^2}{h^2} - (1-au)\frac{c^2}{h^2}.$$

Differentiating with respect to ϕ yields

$$\frac{d^2u}{d\phi^2} + u = \frac{c^2a}{2k^2} + \frac{3au^2}{2}$$

or, substituting for a using (A.94),

$$\frac{d^2u}{d\phi^2} + u = \frac{\gamma m}{h^2} + \frac{3\gamma m u^2}{c^2}. \tag{A.102}$$

This is the usual Newtonian equation with the small 'correction' term $3\gamma m u^2/c^2$ added. The general solution to the Newtonian equation is

$$u = \frac{\gamma m}{h^2}(1 + e\cos(\phi - \phi_o))$$

where e and ϕ_o are constants of integration. Hence, we assume that the general solution to (A.102) is of the form

$$u = \frac{\gamma m}{h^2}(1 + e\cos(\phi - \phi_o)) + \frac{\gamma m}{c^2}u'$$

where the last term represents a small correction to the Newtonian result. Substituting into (A.102) and working to first order in the small parameter $\gamma m/c^2$ yields

$$\frac{d^2u'}{d\phi^2} + u' = \frac{3\gamma^2m^2}{h^4}\Big[1 + 2e\cos(\phi - \phi_o) + e^2\cos^2(\phi - \phi_o)\Big].$$

A particular integral of this equation is

$$u' = \frac{3\gamma^2m^2}{h^4}\Big[1 + e\phi\sin(\phi - \phi_o) - \frac{e^2}{3}cos^2(\phi - \phi_o) + \frac{e^2}{3}\Big].$$

The only term amongst these 'corrections' to the Newtonian orbit which is observationally significant is the second since it is additive in ϕ, i.e. the error accumulates as one observes the planet moving around it orbit many times. Hence, neglecting *all* other terms

$$u = \frac{\gamma m}{h^2}\Big[1 + e\cos(\phi - \phi_o) + \frac{3\gamma^2m^2}{h^2c^2}e\phi\sin\phi - \phi_o\Big].$$

Since $\gamma m/c^2$ is small this can be written as

$$\begin{aligned} u &= \frac{\gamma m}{h^2}\Big[1 + e\Big(\cos\frac{3\gamma^2m^2}{h^2c^2}\phi\cos(\phi - \phi_o) + \sin\frac{3\gamma^2m^2}{h^2c^2}\phi\sin\phi - \phi_o\Big)\Big] \\ &= \frac{\gamma m}{h^2}\Big[1 + e\cos\Big(\phi\Big\{1 - \frac{3\gamma^2m^2}{h^2c^2}\Big\} - \phi_o\Big)\Big]. \end{aligned}$$

This orbit *looks* like an ellipse but the period is now

$$2\pi\left(1-\frac{3\gamma^2 m}{h^2c^2}\right)^{-1} \simeq 2\pi\left(1+\frac{3\gamma m^2}{h^2c^2}\right).$$

From this point of view the orbit is not closed and is a precessing ellipse whose major axis is rotating through an angle $6\pi\gamma m^2/h^2c^2$ radians per revolution of the planet. In the case of the planet Mercury the angle of rotation is $43''$/century which agrees well with observational results.

A.3.5 *The deflexion of light*

Consider a light ray passing through the gravitational field in the neighbourhood of the sun (see Fig. A.1).

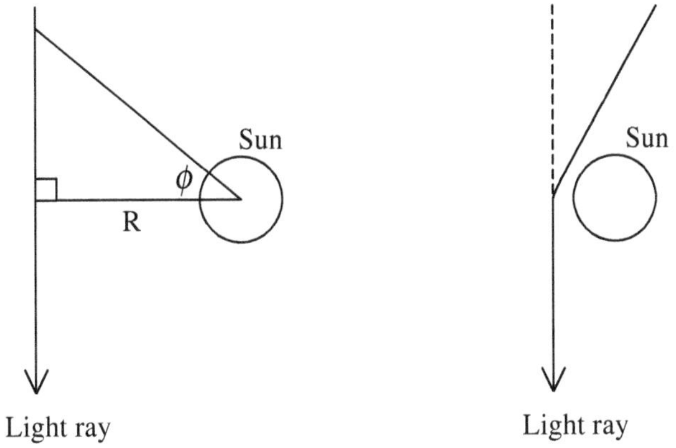

Fig. A.1 Deflexion of light ray near the sun

The field is described by the Schwarzschild solution and the path of the light ray will be a null geodesic. The equations of the null geodesics are analogous to the equations (A.99)–(A.101) with the dot now denoting differentiation with respect to some affine parameter and with the right-hand side of (A.101) zero. The equation analogous to (A.102) now becomes

$$\frac{d^2u}{d\phi^2}+u=\frac{3\gamma m}{c^2}u^2. \tag{A.103}$$

The general solution to the corresponding Newtonian equation

$$\frac{d^2u}{d\phi^2}+u=0$$

is

$$u = A\cos(\phi - \phi_o)$$

where A and ϕ_o are constants of integration. This solution represents a straight line. Hence we assume that the general solution to (A.103) is of the form

$$u = A\cos(\phi - \phi_o) + \frac{\gamma m}{c^2}u'$$

where the last term represents a small correction to the Newtonian result. Substituting into (A.103) and working to the first order in the small parameter $\gamma m/c^2$ yields

$$\frac{d^2u'}{d\phi^2} + u' = 3A^2\cos^2(\phi - \phi_o).$$

A particular integral of this equation is

$$u' = A^2(-\cos^2(\phi - \phi_o) + 2).$$

Hence

$$u = A\cos(\phi - \phi_o) + \frac{\gamma m}{c^2}A^2(-\cos^2(\phi - \phi_o) + 2). \tag{A.104}$$

Now choose the axis $\phi = 0$ to pass through the point at which the ray is closest to the sun and assume this closest distance is R. Then $\phi_o = 0$ and $A = R^{-1}$ so that

$$u = \frac{1}{R}\cos\phi + \frac{\gamma m}{R^2c^2}(-\cos^2\phi + 2).$$

As $r \to \infty, u \to 0$ and the angle ϕ is given by

$$\cos\phi = \frac{R \pm \sqrt{R^2 + 8\gamma^2m^2/c^4}}{2\gamma m/c^2} \simeq \frac{R \pm R(1 + 4\gamma^2m^2/c^4R^2)}{2\gamma m/c^2}.$$

Since $|\cos\phi| \leq 1$ the negative sign must be taken so that

$$\cos\phi \simeq -\frac{2\gamma m}{c^2R} = -\sin\frac{2\gamma m}{c^2R}.$$

It follows that the light ray does not follow a straight line. It approached with an asymptotic angle $-\frac{\pi}{2} - 2\gamma m/c^2R$ and departs with an asymptotic angle $\frac{\pi}{2} + 2\gamma m/c^2R$. The angle through which the light ray is deflected is

$$4\gamma m/c^2R \text{ radians.}$$

For the angle to be detectable R must be as small as possible, that is the light ray should just graze the edge of the sun. The deflexion for such a light ray is $1.75''$. Because of the brilliance of the sun the observational measurement of this deflexion is impossible except during an eclipse. Agreement with observation is poor.

A.3.6 *The gravitational red shift*

Consider an atom radiating photons. The frequency of emission of photons will be taken to be the number of photons emitted per unit 'distance' s rather than per unit time t. With such a definition the frequency is a *scalar* quantity.

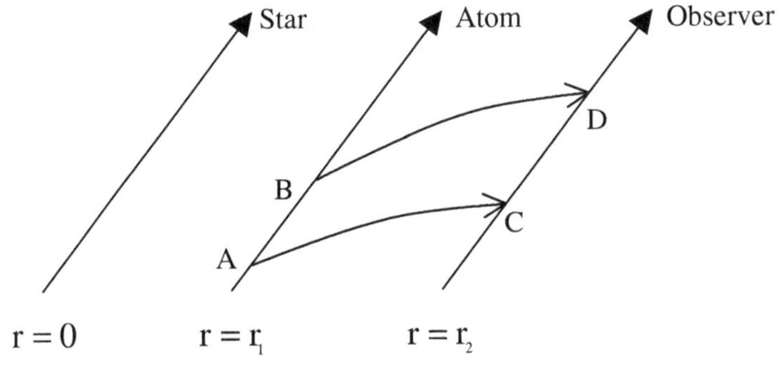

Fig. A.2 Gravitational red shift

Now suppose that the above atom is at rest relative to a star and that the photons are received by an observer also at rest relative to the star. Suppose consecutive photons are emitted at A and B and received at C and D. Then the frequencies of emission and reception are given by

$$\nu_e = \frac{1}{ds_{AB}}$$

and

$$\nu_r = \frac{1}{ds_{CD}}.$$

The gravitational field of the star is described by the Schwarzschild solution. Hence, since the atom and observer are at rest relative to the star, it follows that along their world lines

$$ds^2 = (1 - a/r)dt^2.$$

Hence

$$\nu_e = \frac{1}{(1 - a/r_1)^{\frac{1}{2}}} dt_{AB}$$

and

$$\nu_r = \frac{1}{(1 - a/r_2)^{\frac{1}{2}}} dt_{CD}.$$

Since the Schwarzschild solution is static it may be assumed that $dt_{AB} = dt_{CD}$ and so

$$\frac{\nu_e}{\nu_r} = \frac{\left(1 - a/r_2\right)^{\frac{1}{2}}}{\left(1 - a/r_1\right)^{\frac{1}{2}}}.$$

There is therefore a change in the frequency between emission and reception of the photons and a corresponding shift in the spectral lines of the radiation. The greatest shift occurs when r_2 is a maximum and r_1 a minimum, that is when the atom is on the *surface* of the star and when the observer is at a great distance from the star. Then the proportional increase in frequency is

$$\begin{aligned}\frac{\nu_r - \nu_e}{\nu_r} &= 1 - (1 - a/r_1)^{-\frac{1}{2}} \\ &= 1 - \left(1 - \frac{2\gamma m}{c^2 r_1}\right)^{-\frac{1}{2}} \simeq -\frac{\gamma m}{c^2 r_1}.\end{aligned}$$

Such a shift (to the red) is observed and the expression found here agrees well with observations of Sirius B but not so well with observations of the sun.

A.4 Exercises

A.1 Consider a plane $lx+my+uz = p$. Write down a vector $\mathbf{n}$ perpendicular to the plane. Prove that the coordinates x and y can be chosen as parameters on the plane provided that the plane is *not* perpendicular to the xy plane. Find the natural basis vectors associated with these coordinates and show that they are parallel to the plane. Deduce that the tangent plane at any point P_o coincides with the plane itself.

A.2 (r, θ, ϕ) are spherical polar coordinates. A sphere S of radius a, centred at the origin, is parameterised by θ and ϕ. Draw a picture illustrating the two families of parametric curves. Find the corresponding natural basis vectors for $T_{P_o}(S)$, where P_o is the point having spherical polar coordinates $(a, \frac{\pi}{4}, \frac{\pi}{4})$. Verify that both basis vectors are orthogonal to the radius vector through P_o.

A.3 Three tensors having components $T^i{}_{jk}, T_{ijk}$ and T^{ijk} respectively are defined at a point P_o of a given manifold. To what vector spaces do these tensors belong?

A.4 Write down the transformation laws for the components of the tensors in Exercise A.3.

A.5 The Kronecker delta δ^i_j is defined in any coordinate system by $\delta^i_j = 1$ if i and j take the same value and $= 0$ if i and j take different values. Show, by verifying that the appropriate transformation law is satisfied, that δ^i_j are the components of a tensor belonging to $\in T_{P_o}{}^{(1)}_{(1)}(M)$. Would δ^i_j, defined in a similar way, be the components of a tensor $\in T_{P_o}{}^{(0)}_{(1)}(M)$?

A.6 Calculate the number of independent components n_1, n_2 and n_3 of a general, a symmetrical and a skew symmetrical tensor of covariant valence two (work in either an n or a four-dimensional manifold) respectively. Verify that

$$n_1 = n_2 + n_3.$$

Repeat the calculation for tensors of covariant valence three and deduce that a general such tensor cannot be written as the sum of two tensors, one of which is completely symmetrical, the other completely skew symmetrical.

A.7 The components of the tensor T satisfy the equations $T^{(ij)k} = 0$ and $T^{i[jk]} = 0$. Prove that T is identically zero.

A.8 Prove that the non-singularity of a tensor with components $T^i{}_j$ is invariant under all coordinate transformations ($T^i{}_j$ is non-singular if and only if $|T^i{}_j| \neq 0$).

A.9 Prove that, if δ^i_j is the Kronecker delta, then

$$v_i \delta^i_j = v_j.$$

Use the second quotient theorem to deduce that δ^i_j is a tensor.

A.10 S_{ij} and A^{ij} are the components of a symmetric and a skew symmetric tensor respectively. Prove that the contracted product

$$S_{ij}A^{ij}$$

is identically zero.

A.11 The array $\hat{S}^i{}_{jk}$ is defined in each coordinate system, the contracted product of $\hat{S}^i{}_{jk}$ with the components of an arbitrary tensor being the components of a contravariant vector. Prove that $\hat{S}^i{}_{jk}$ are themselves the components of a unique tensor (all quantities are defined at some point $P_o \in M$).

A.12 The contracted product $T_{ij}v^iv^j$ defined at a point $P_o \in M$ is a scalar for all vectors v^i. Show that T_{ij} are the components of a unique tensor. Can anything be deduced about $T_{[ij]}$?

A.13 Prove that $v_{[i,j]}$ defines a unique tensor for each vector having components v_i.

A.14 Obtain the transformation law for the Christoffel symbol Γ^i_{jk} under a change of coordinates.

A.15 Write down an expression for $T^i{}_{pq;j}$.

A.16 A certain Riemannian geometry is defined by

$$ds^2 = (l - r^2)dt^2 - dr^2(1 - r^2)^{-1} - r^2 d\theta^2 - r^2 \sin^2\theta d\phi^2.$$

Putting $(x^1, x^2, x^3, x^0) \equiv (r, \theta, \phi, t)$ write down the components g_{ij} of the metric tensor g and deduce expressions for the corresponding contravariant components g^{ij}.

A.17 In a two-dimensional Riemann space with metric

$$ds^2 = du^2 + 2\lambda du dv + dv^2$$

where λ is a function of u and v, show that the tangent vectors to the curves $u =$ constant form a field of parallel vectors along the curves $v =$ constant. [Definition: v^i is a field of parallel vectors along u^i if and only if $v^i{}_{;j}u^j \propto v^i$].

A.18 With the usual definition of covariant derivatives, the tensor $R^p{}_{qrs}$ is defined by

$$A_{q;rs} - A_{q;sr} = R^p{}_{qrs}A_p$$

where A_p is an arbitrary vector.

(a) By taking $A_q = C_p B^p{}_q$, where C_p and $B^p{}_q$ are an arbitrary vector and tensor, prove that

$$B^p{}_{q;rs} - B^p{}_{q;sr} = R^t{}_{qrs} B^p{}_t - R^p{}_{trs} B^t{}_q.$$

(b) By now taking $B^p{}_q = A^p{}_{;q}$ and assuming the usual symmetries of $R^p{}_{qrs}$, establish the Bianchi identities

$$R^p{}_{qrs;t} + R^p{}_{qst;r} + R^p{}_{qtr;s} = 0.$$

A.19 Show that the equation of a timelike geodesic can be obtained from the variational principle

$$\delta \int_A^B ds = \delta \int_A^B (g_{ij} \dot{x}^i \dot{x}^j)^{\frac{1}{2}} d\lambda = 0.$$

A.20 Verify that (A.66) is indeed identical to (A.64) with $\alpha = 0$.

A.21 Show that the null geodesics are unaltered if the metric tensor g is replaced by γg where γ is a function of the coordinates.

A.22 Use the Euler–Lagrange equations to obtain the geodesic equations of the Riemann space of Exercise A.17 and deduce expressions for the Christoffel symbols.

A.23 The Ricci tensor of a certain space-time is given by

$$R_{ij} = \phi g_{ij}.$$

Using the contracted Bianchi identities prove that ϕ is constant.

Bibliography

Aharanov, Y. and Susskind, L. (1967). Observability of the sign change of spinors under 2π rotations, *Phys. Rev.* **158**, pp.1237–1238.

Ahlfors, L.V. and Sario, L. (1960). *Riemann Surfaces* (Princeton University Press, Princeton).

Bade, W.L. and Jehle, H. (1953). An introduction to spinors, *Rev. Mod. Phys.* **25**, pp.714–28.

Bampi, F. and Caviglia, G. (1983). Third-order tensor potentials for the Riemann and Weyl tensors, *Gen. Rel. Grav.* **15**, pp.375–386.

Bergmann, P.G. (1957). Two-component spinors in general relativity, *Phys. Rev.* **107**, pp.624–29.

Brauer, R. and Weyl, H. (1935). Spinors in n dimensions, *Am. J. Math.* **57**, pp.425–49.

Buchdahl, H.A. (1959). On extended conformal transformations of spinors and spinor equations, *Nuovo Cim.* **11**, pp.496–506.

Carmeli, M. and Malin, S. (2000). *An Introduction: Theory of Spinors* (World Scientific, Singapore).

Cartan, É. (1966). *The Theory of Spinors* (Hermann, Paris).

Chandrasekhar, S. (1979). An introduction to the theory of the Kerr metric and its perturbations, in *General Relativity: An Einstein Centenary Survey*, ed. S.W. Hawking and W.Israel (Cambridge University Press, Cambridge).

Chevalley, C. (1954). *The Algebraic Theory of Spinors* (Columbia University Press, New York).

Corson, E.M. (1953). *Introduction to Tensors, Spinors and Relativistic Wave Equations* (Blackie, Glasgow).

Cvitanovic, P. (1976). Group theory for Feynman diagrams in non-abelian gauge theories, *Phys. Rev.* **D14**, pp.1536–1553.

Cvitanovic, P. and Kennedy, A.D. (1982). Spinors in negative dimensions, *Phys. Scripta.* **26**, pp.5–14.

de Parga, G.A., Chavoya, O.A., and López–Bonilla, J.L. (1989). Lanczos potential, *J. Math. Phys.* **8**, pp.1294–1295.

Dodson, C.T.J. and Poston, T. (1977). *Tensor Geometry* (Pitman, London).

Dolan, P. and Kim, C.W. (1994). Some solutions of the Lanczos vacuum wave equation, *Proc. Roy. Soc. Lon.* **A447**, pp.577–585.

Edgar, S.B. (1987). *Gen. Rel. Grav.* **19**, 1149.

Edgar, S.B. (1994). Nonexistence of the Lanczos potential for the Riemann tensor in higher dimensions, *Gen. Rel. Grav.* **26**, pp.329–332.

Edgar, S.B. (2001). Existence of Lanczos potentials and superpotentials for the Weyl spinor/tensor, *Class. Quantum. Grav.* **18**, pp.2297–2304.

Ehlers, J. (1974). The geometry of the (modified) GHP-formalism, *Commun. Math. Phys.* **37**, pp.327–329.

Einstein, A. (1916). Die Grundlage der allegemeinen Relativitätstheorie, *Ann. Phys.* **49**, pp.769–822.

Einstein, A. (1926). Über die formale Beziehung des Riemannshen Krummungstensors zu den Feldgleichungen der Gravitation, *Maths. Ann.* **97**, pp.99–108.

Flanders, H. (1963). *Differential Forms* (Academic Press, New York).

Geroch, R. (1968). Spinor structure of space-times in general relativity I, *J. Math. Phys.* **9**, pp.1739–1744.

Geroch, R. (1970). Spinor structure of space-times in general relativity II, *J. Math. Phys.* **11**, pp. 343–348.

Geroch, R., Held, A. and Penrose, R. (1973). A space-time calculus based on pairs of null directions, *J. Math. Phys.* **14**, pp.874–881.

Goldberg, J.N. and Sachs, R.K. (1962). A Theorem on Petrov Types, *Acta Phys. Polon.* **22**, p.13–32.

Hansen, R.O., Janis, A.I., Newman, E.T., Porter, J.R. and Winicour, J. (1976). Tensors, spinors and functions on the unit sphere, *Gen. Rel. Grav.* **7**, pp.687–693.

Held, A. (1974). A formalism for the investigation of algebraically special metrics I, *Com. Math. Phys.* **37**, pp.311–326.

Held, A. (1975). *Ditto II. Com. Math. Phys.* **44**, pp.211–222.

Huggett, S.A. and Tod, K.P. (1985). *An Introduction to Twistor Theory*, London Math. Soc. Lecture Notes Series, (Cambridge University Press, Cambridge).

Hulse, R.A. and Taylor, J.H. (1975). Discovery of a pulsar in a binary system, *Astrophys. J.* **195**, pp.51–53.

Illge, R. (1988). On potentials for several classes of spinor and tensor fields in curved spacetimes, *Gen. Rel. Grav.* **20**, pp.551–564.

Infeld, L. and van der Waerden, B.L. (1933). Die Wellungleichung des Elektrons in der allgemeinen Relativitätstheorie, *Sitz. Ber. Preuss. Akad. Wiss. Physik.-Math.*, **K1, 9**, pp.380–401.

Kasner, E. (1921). Geometrical theorems on Einstein's cosmological equations, *Am. J. Math.* **43**, pp.217–221.

Kinnersley, W. (1969). Type D vacuum metrics *J. Math. Phys.* **10**, 1195.

Kramer, D., Stephani, H., MacCallum, M.A.H. and Herlt, E. (1980). *Exact Solutions of Einstein's Field Equations* (Cambridge University Press, Cambridge).

Lanczos, C. (1938). A remarkable property of the Riemann–Christoffel tensor in four dimensions, *Ann. Math.* **39**, pp.842–850.

Lanczos, C. (1949). Lagrangian multiplier and Riemann spaces, *Rev. Mod. Phys.* **21**, pp.497–502.

Lanczos, C. (1962). The splitting of the Riemann tensor, *Rev. Mod. Phys.* **34**, pp.379–389.

Laporte, O. and Uhlenbeck, G.E. (1931). Application of spinor analysis to the Maxwell and Dirac equations, *Phys. Rev.* **37**, pp.1380–1552.

Maher, W.F. and Zund, J.D. (1968). A spinor approach to the Lanczos spin tensor, *Il Nuovo Cimento* **57A**, pp.638–649.

Massa, E. and Pagani, E. (1984). Is the Riemann tensor derivable from a tensor potential?, *Gen. Rel. Grav.* **16**, pp.805–816.

Misner, C.W., Thorne, K.S. and Wheeler, J.A. (1973). *Gravitation* (W.H. Freeman, San Francisco).

Morrow, J. and Kodaira, K. (1971). *Complex Manifolds* (Holt, Rinehart and Winston, New York).

Newman, E.T. and Penrose, R. (1962). An approach to gravitational radiation by a method of spin coefficients, *J. Math. Phys.* **3**, pp.566–578; Errata: *ditto. Ibid.* **4**, p.998.

Novello, M. and Velloso, A.L. (1987). The connection between general observers and Lanczos potential, *Gen. Rel. Grav.* **19**, pp.1251–1265.

O'Donnell, P. (1997). Analysis of the Lanczos tensor incorporating generating techniques for some empty spacetimes, University of Sussex DPhil. thesis.

Payne, W.T. (1952). Elementary spinor theory, *Am. J. Phys.* **20**, pp.253–262.

Penrose, R. (1959). The apparent shape of a relativistically moving sphere, *Proc. Camb. Phil. Soc.* **55**, pp.137–139.

Penrose, R. (1960). A spinor approach to general relativity, *Ann. Phys.* **10**, pp.171–201.

Penrose, R. (1962). General relativity in spinor form, in *Les Théories Relativistes de la Gravitation*, eds. A. Lichnerowicz and M.A. Tonnelat (CNRS, Paris).

Penrose, R. (1968). Structure of space-time, in *Battelle Rencontres, 1967 Lectures in Mathematics and Physics*, eds. C.M. DeWitt and J.A. Wheeler (Benjamin, New York).

Penrose, R. (1971). Application of negative dimensional tensors, in *Combinatorial Mathematics and its Applications*, ed. D.J.A. Welch (Academic Press, London).

Penrose, R. (1972b). Spinor classification of energy tensors, *Gravitatsiya Nauk dumka*, Kiev, pp.203–215.

Penrose, R. (1983). Spinors and torsion in general relativity, *Found. Phys.* **13**, pp.325–340.

Penrose, R. and Rindler, W. (1984). *Spinors and Space-Time Vol.1: Two-Spinor Calculus and Relativistic Fields* (Cambridge University Press, Cambridge).

Penrose, R. and Rindler, W. (1986). *Spinors and Space-Time Vol.2: Spinor and Twistor Methods in Space-Time Geometry* (Cambridge University Press, Cambridge).

Pirani, F.A.E. and Schild, A. (1961). Geometrical and physical interpretation of the Weyl conformal curvature tensor, *Bull. Acad. Pol. Sci.*, Ser. Sci. Math. Astron. Phys. **9**, pp.543–547.

Rainich, G.Y. (1925). Electrodynamics in the G.R. theory, *Trans. Am. Math. Soc.* **27**, p.106–136.

Rindler, W. (1966). What are spinors? *Am. J. Phys.* **34**, pp.937–942.

Schild, A. (1967). Lectures on general relativity theory, in *Relativity Theory and Astrophysics, I. Relativity and Cosmology*, ed. J.Ehlers (Amer. Math. Soc. Providence).

Stewart, J.M. and Walker, M. (1974). Perturbations of space-times in general relativity, *Proc. Roy. Soc. London* **A341**, pp.49–74.

Takeno, H. (1964). On the spintensor of Lanczos, *Tensor* **15**, pp.103–119.

Trautman, A., Pirani, F.A.E. and Bondi, H. (1965). *Lectures on General Relativity* (Prentice-Hall, Englewood Cliffs).

Udeschini, E.B. (1977). Su Un Tensore Triplo 'Potenziale' del Tensore Di Riemann, Parte I, *Rendiconti Istituto Lombardo, A***111**, pp.466–475.

Udeschini, E.B. (1978). Su Un Tensore Triplo 'Potenziale' del Tensore Di Riemann, Parte II, *Rendiconti Istituto Lombardo, A***112**, pp.110–122.

Udeschini, E.B. (1980). A natural scalar field in the Einstein gravitational theory, *Gen. Rel. Grav.* **12**, pp.429–437.

Udeschini, E.B. (1981). On the relativistic extension of the equations of the classical gravitational potential, *Meccanica* **16**, pp.75–79.

van der Waerden, B.L. (1929). Spinoranalyse, *Nachr. Akad. Wiss. Götting Math.-Physik* **K1**, pp.100–109.

Veblen, O. (1933a). Geometry of two-component spinors, *Proc. Nat. Acad. Sci.* **19**, pp.462–474.

Veblen, O. (1933b). Geometry of four-component spinors, *Proc. Nat. Acad. Sci.* **19**, pp.503–517.

Veblen, O. and Taub, A.H. (1934). Projective differentiation of spinors, *Nat. Acad. Sc. Proc.* **20**, pp.85–92.

Veblen, O. (1934). Spinors, *Science* **80**, pp.415–419.

Veblen, O. and von Neumann, J. (1936). *Geometry of Complex Domains*, mimeographed notes, issued by the Institute for Advanced Study Princeton (reissued 1955).

Wells, R.O. (1973). *Differential Analysis in Complex Manifolds* (Prentice-Hall, Englewood Cliffs).

Whittaker, E.T. (1937). On the relations of the tensor-calculus to the spinor-calculus, *Proc. Roy. Soc. London* **A158**, pp.38–46.

Zund, J.D. (1975). The theory of the Lanczos spinor, *Ann. di Mat. Pura ed Appl.* **109**, pp.239–268.

Index

www.ingramcontent.com/pod-product-compliance
Lightning Source LLC
Chambersburg PA
CBHW050640190326
41458CB00008B/2360
* 9 7 8 9 8 1 2 3 8 3 0 7 5 *